Urban Green Spaces

Edited by Rui Alexandre Castanho
and José Cabezas Fernández

Published in London, United Kingdom

Urban Green Spaces
http://dx.doi.org/10.5772/intechopen.97910
Edited by Rui Alexandre Castanho and José Cabezas Fernández

Contributors
Vasco Barbosa, Mónica Marcela Suárez Pradilla, Juan Eduardo Chica-Mejía, Jairo Eduardo Galvis Bonilla, Cristina Santos, Cristina M. Monteiro, Prashanti Rao, Somaina Islary, Kapil Natawadkar, Akanksha Sangwan, Nand Kumar, Ashwani Kumar, Anjali Saraswat, Satish Pipralia, Ruffina Thilakaratne, Jaran R. Wood, Kim Rosembom, Başak Kılıç Taşeli, Rui Alexandre Castanho, Teresa Batista, Salustiano Torre, José Fermoso, Nuno Morais, Ramona Avila, José Cabezas Fernández, Carlos Pinto Gomes

First published in London, United Kingdom, 2022 by IntechOpen
IntechOpen is the global imprint of INTECHOPEN LIMITED, registered in England and Wales, registration number: 11086078, 5 Princes Gate Court, London, SW7 2QJ, United Kingdom

British Library Cataloguing-in-Publication Data
A catalogue record for this book is available from the British Library

Additional hard and PDF copies can be obtained from orders@intechopen.com

Urban Green Spaces
Edited by Rui Alexandre Castanho and José Cabezas Fernández
p. cm.
Print ISBN 978-1-80355-156-2
Online ISBN 978-1-80355-157-9
eBook (PDF) ISBN 978-1-80355-158-6

Meet the editor

Rui Alexandre Castanho has a Ph.D. in Sustainable Planning in Borderlands. He is a professor at WSB University, Poland, and a visiting professor at the University of Johannesburg, South Africa. Dr. Castanho collaborates with several research groups all over the world.

José Cabezas Fernández has a Ph.D. in Ecology and has extensive research experience in sustainable management, conservation, and recovery of natural ecosystems (water, soil, fauna, flora, etc.), land use planning, environmental impact, remote sensing and GIS, light detection and ranging (LIDAR), and spatial data infrastructure (SDI), aspects that are fundamentally oriented towards the conservation of transboundary Mediterranean ecosystems and their sustainability. His projects have focused on the development of geographic models of environmental and territorial management for rural areas with low population density for better and more coherent management of the territory.

Contents

Preface

Throughout history, anthropic activities have resulted in deep modifications in land use and territory occupation. In fact, most of the world's original landscapes have experienced substantial transformations by anthropic actions. This issue results from a combination of factors such as demographic growth and economic structure changes from the primary sector to the industrial and services sector. As the world's cities have grown, the relevance of urban green spaces has become more apparent, considering all the benefits they can provide to improve our urban living standards.

This book, which contains nine chapters in two sections, is a powerful scientific contribution to help us plan our cities and the associated urban green spaces as well as our common future.

In this regard, the book starts with an introductory chapter (Chapter 1), where a brief description of the benefits that urban green spaces could deliver us and, at the same time, the obstacles the urban planners to faces to create the newest solutions for incorporating these green areas within urban territories.

Chapter 2 by Sangwan et al. analyzes the structural nature of urban green spaces and their benefits from an Indian perspective, proposing strategic measures for urban planners.

Chapter 3 by Rao et al. examines how integrating green areas in educational centers contributes to improving mental health and cognitive development, for both students and employees at these centers.

Chapter 4 by Barbosa et al. applies remote sensing techniques to a South American city to analyze its urban expansion, evidencing a disconnection and disarticulation between public spaces and green areas, with a total absence of green corridors.

Despite the spectacular nature of large parks, small green spaces play a fundamental role in the well-being of humans in high-density cities. Chapter 5 by Thilakaratne discusses this premise in the context of Hong Kong, delving into the parameters necessary to design these mini green areas.

Chapter 6 by Başak Kılıç Taşeli analyzes stormwater management in urban areas and the problems caused by inefficient management systems, proposing the application of green infrastructures to preserve runoff water quality.

Chapter 7 by Batista et al. examines the adoption of nature-based solutions for "green" building coverings to improve the resilience of public buildings using native vegetation.

Chapter 8 by Santos et al. studies green roofs and their effect on rainwater, analyzing their effects on water retention and the quality of their properties.

Finally, Chapter 9 by Cristina Santos and Dr. Cristina M. Monteiro tells us how nature-based solutions through the use of new solutions could improve urban sustainability as well as support stormwater management.

Rui Alexandre Castanho
WSB University,
Dabrowa Górnicza, Poland

VALORIZA,
Polytechnic Institute of Portalegre,
Portugal

José Cabezas Fernández
VALORIZA,
Polytechnic Institute of Portalegre,
Portugal

University of Extremadura,
Badajoz, Spain

Section 1

Starting Point: Benefits and Functionalities of Urban Green Spaces

Chapter 1

Introductory Chapter: Urban Green Spaces – An Opening Framework

Rui Alexandre Castanho and José Cabezas Fernández

1. Introduction

Throughout the evolution of human populations, there have been profound transformations in the occupation of the territory. The vast majority of the original landscapes of our planet have undergone significant modifications by human activity. More specifically, in Europe, the hunter-gatherer stage they have led to the disappearance of primeval forests. These have been replaced, in the areas where greater control was exercised, by intensive agricultural territories and large cities, in the longer term [1–3].

Initially, the type of distribution was predominantly rural, with low density and small cities with poor communications. In recent decades, we have witnessed a change in the forms of land occupation, with an increase in artificial surfaces. This situation is a consequence of two factors, demographic growth, and the change in economic structures from the primary sector to the industrial and services sector [4]. Cities have suffered population growth caused by multiple causes, from the typical concentration of industry and improvement in the provision of public services in large cities, depletion of mining areas, and even the effect of natural disasters such as earthquakes that caused the exodus almost instantaneously among entire populations [5, 6].

However, unlike what happened at the beginning of these change processes, urbanization model that consumes much more land is taking place lately. This low-density urban development system is caused by increased purchasing power, greater mobility caused by motorization, and an explosion of mass tourism. In recent times, it is the tourism of cities that can act as a focus of attraction for migration due to the more excellent value that people give to the quality of life and apply it to their free time [7]. In fact, a panorama affects the natural and agricultural landscape due to the dense fabric of the urban and its connection routes. This situation is aggravated by demographic perspectives, as shown in the study by UN-DESA [4], which indicates that in 1950 only 30% of the world's population lived in cities, while in 2018, that percentage rose to 55%.

In all these contexts, it would be necessary to consider what is urban and rural. The latter would be those that do not have urban characteristics or, using rural identity criteria, those with a significant presence of green or open spaces [8, 9]. If we resort to European terminology, the urban/rural typology is based on two dimensions: the degree of human intervention and urban influence [10]. The first is represented by the proportion of artificial surfaces (industry, housing, and communication routes, among several other examples). At the same time, the second has to do with the range of the primary urban center in each area and its population density. Consequently, current trends in land use, from rural to urban, have two aspects.

On the one hand, there is its contribution to increasing social and economic benefits. Nevertheless, on the other, they lead, in the longer term, to a series of alterations with adverse effects on the functioning of ecosystems. This disorganized and chaotic development was later attempted to be corrected with urban planning procedures, whose objective was to eliminate urban chaos and prevent it from happening again [11, 12].

This does not mean these negative consequences cannot be stopped or improved. Thus, strategies such as an increase in agricultural efficiency, the brake on erosion, sustainable agroforestry policies (such as the reintroduction of livestock in abandoned areas to curb the proliferation of forest fires), and the increase of green spaces in urban areas could be applied.

This last aspect, urban green spaces, is of considerable importance as it is related to some of the environmental problems of cities. Until the eighteenth century and since Roman times, the formal and physical relationship between the city and its surroundings remained relatively unchanged, with a closed city considered the central factor and an environment without too many alterations. The walls also included social connotations, the dominant classes lived inside, and the rest, people of lower social status, lived or worked in the surroundings. This thinking evolved into the concept that something valuable should be locked up, leading to the development of private gardens for the enjoyment of a few [13, 14]. The city has recently evolved toward the need to develop an urban planning concept in which the existence of green spaces has become more critical. Although new urban planning models were introduced in the 1920s, environmental issues were unimportant. However, at the beginning of urban theories, the environment was considered essential for balance and quality of life [15] and especially in the works of McHarg [16], who sought to bridge the gap between ecology and land planning and gave way to an ecological approach to planning. In recent times, Steiner [17] indicates that: *"(...) planning is a philosophy for organizing actions that enable people to predict and visualize the future of any land area."*

The processes of territorial expansion in urban fabrics have caused a significant increase in the generation of superheated air, higher concentrations of pollution in the atmosphere, and a scarce presence of vegetation. This set of factors contributes to modifying local climatic conditions, giving rise to urban heat islands. In addition, the waterproofing of a large part of the territory's surface hinders the infiltration of precipitation, reducing the absorption rate of the soil. All these, together with the increasingly frequent episodes of torrential rains, increase the risk of flooding in urban areas [18, 19].

Within green areas, trees have essential effects on the environmental conditions of cities. Concerning contamination by fine particles in the atmosphere, they contribute to buffering it since they behave as a sink for these particles through the mechanism of variation in stomatal conductance and photosynthetic rate [20]. Gaseous pollutants from the atmosphere are also filtered by the trees in cities, as is the case with NOX emissions caused by vehicles traveling on communication routes [21, 22].

Trees' shade effect is recognized and widely studied, highlighting its participation in buffering the urban heat island [23–25]. The reduction in the incidence of radiation and the increase in humidity due to perspiration are decisive factors in these processes. In a recent study, in European cities, it has been seen that when comparing treeless areas with the presence of trees, significant temperature drops were produced, more pronounced in the north than in the Mediterranean area and the Iberian Peninsula [26]. The importance of green infrastructure, its diffusion, and increase acquires excellent value when considering the emergencies and deaths causing heat waves throughout the planet [27–29].

Therefore, the above-mentioned links to the concept of ecosystem services. In recent years, much attention has been paid to ecosystems because of the services that allow life reside in them. In the middle of the last century, it was believed that the loss of habitat, biological species, and the depletion of natural resources was due, to a large extent, to the fact that the services provided by ecosystems did not have a monetary value, focusing on values associated with cultural services [30]. Subsequently, attempts were made to price environmental goods and services [31, 32]. However, certain services are intangible (pollination, climate regulation, and biodiversity are just a few examples.), and their value is ecological and not monetary. It was in 1981 when the idea of giving value to the services provided by nature from both an economic and a practical point of view crystallized [33]. The concept of ecosystem services was developed to draw attention on the benefits that ecosystems provide to society. It evolved to become an essential framework for promoting and designing environmental management. The valuation of ecosystem services was originally designed for nonurban systems [34]; hence, the need to create new models that reflect those provided by urbanized environments [35, 36].

Despite what ecosystem services represent, it is necessary to consider that the physical surface of the city itself is not enough to develop life skills in cities. The needs far exceed the size they have. Hence, concepts such as the ecological footprint have been developed [37].

With postmodernism emerging in the second half of the twentieth century, a "new urbanism" appeared, bringing the awakening of urban ecology and environmental concerns that filled the last treatment gap in social and environmental-systemic issues [38].

Summing up the above statements, analyzing green spaces from the perspective of Urban Ecology should be considered. This subdiscipline of ecology emerged in the last decades of the twentieth century. It uses and integrates terminologies and methodologies from various disciplines such as ecology, land planning, architecture, economics, political science, geography, sociology, engineering, anthropology, and health sciences [39]. Humans are part of ecosystems, but instead of just another organism, we are a force of nature that acts to modify our environment [39]. People can relate well and create social groups capable of influencing the administrative bodies that make environmental decisions. In other words, politics or sociology influences urban ecosystems depending on our culture [40]. In this way, urban ecology is helping to make cities friendlier.

Despite all these advantages that green areas present, care must be taken and not take steps back, as in the USA with the loss of tree canopy (7000 hectares per year) between 2009 and 2014 [41].

2. Final Remarks

As final remarks, urban green areas, ecosystem services, ecological footprint, and urban ecology constitute a chain that links the problems and solutions that must govern the management of urban areas and that administrations must incorporate. However, time is running out against all approaches. Compared with the rural world, cities' socioeconomic benefits will continue to displace the population in larger urban areas. The policies to, if not stop, at least mitigate these migrations go through finding ways to fix the population in these rural areas, the eternal "philosopher's stone" of territorial planning. Nevertheless, what about existing cities and those that will continue to grow? Among all the applicable measures, what contribution can green areas provide? These are the questions that, in part, try to clarify the works that professionals from very diverse disciplines worldwide are trying to answer and to which this book aims to be a small contribution.

Author details

Rui Alexandre Castanho[1,2*] and José Cabezas Fernández[2,3]

1 WSB University, Dabrowa Górnicza, Poland

2 VALORIZA, Polytechnic Institute of Portalegre, Portugal

3 University of Extremadura, Badajoz, Spain

*Address all correspondence to: acastanho@wsb.edu.pl

References

[1] González-Bernáldez F. Ecología y Paisaje. Madrid, Spain: Blume Ediciones; 1981

[2] Portney K. Taking Sustainable Cities Seriously: Economic Development, the Environment, and Quality of Life in American Cities. Cambridge, UK: MIT Press; 2003

[3] Etter A, McAlpine CA, Seabrook L, Wilson KA. Incorporating temporality and biophysical vulnerability to quantify the human spatial footprint on ecosystems. Biological Conservation. 2011;**144**(5):1585-1594. DOI: org/10. 1016/j.biocon.2011.02.004

[4] UN-DESA, United Nations, Department of Economic and Social Affairs, Population Division. *The World's Cities in 2018—Data Booklet* (ST/ESA/SER.A/417). New York, USA: The United Nations Bookshop; 2018

[5] Marini MB, Mooney PH. Rural economies. In: Cloke P, Marsden T, Mooney PH, editors. Handbook of Rural Studies. London: Sage; 2006. pp. 91-103

[6] Di Figlia L. Turnaround: Abandoned villages, from discarded elements of modern Italian society to possible resources. International Planning Studies. 2016;**21**(3):278-297. DOI: 10.1080/13563475.2016.1186530

[7] Niedomysl T. Tourism and interregional migration in Sweden: An explorative approach. Population, Space and Place. 2005;**11**(3):187-204. DOI: 10.1002/psp.366

[8] Isserman AM. In the national interest: Defining rural and urban correctly in research and public policy. International Regional Science Review. 2005;**28**(4):465-499

[9] Jonard F, Lambotte M, Ramos F, Terres J, Bamps C. Delimitations of Rural Areas in Europe Using Criteria of Population Density, Remoteness and Land Cover. EUR 23757 EN. Luxembourg (Luxembourg): OPOCE. JRC49927; 2009

[10] ESPON. Urban-rural relations in Europe. Bengs C, & Schmidt-Thomé K. eds. ESPON 2006 Programme, Project 1.1.2. Final Report. ESPON Monitoring Commietee, Luxembourg; 2006

[11] Terán F. Historia del Urbanismo en España – Siglos XIX y XX. Madrid, Spain: Cátedra; 1999

[12] Casares S. Ordenación del Território I – El Planeamiento Urbano. University of Oviedo, Oviedo; 2009

[13] Accati E, Gullino P. Plants in gardens and parks of XVIII century in Piedmont. Acta Horticulturae. 2010;**881**:939-944. DOI: 10.17660/ActaHortic.2010.881.157

[14] Fornaris A, Devecchi M, Larcher F. A new approach for the valorisation of the historical garden heritage: The case of the Biella province in Italy. Acta Horticulturae. 2013;**999**:153-157. DOI: 10.17660/ActaHortic.2013.999.21

[15] Howard E. Garden Cities of Tomorrow: A Peaceful Path to Real Reform. London, UK: Swan Sonnenschein & Co.; 1903

[16] McHarg IL. Design with Nature. New York, USA: John Wiley & Sons, Inc.; 1967

[17] Steiner FR. The Living Landscape. An Ecological Approach to Landscape

Planning. New York: McGraw-Hill, Inc.; 2000. p. 2000

[18] Luna T, Rocha A, Carvalho AC, Ferreira JA, Sousa J. Modelling the extreme precipitation event over Madeira Island on 20 February 2010. Natural Hazards and Earth System Sciences. 2011;**11**:2437-2452

[19] Lousada S, Cabezas J, Castanho RA, Gómez JMN. Hydraulic planning in insular urban territories: The case of Madeira Island—Ribeira Brava. Water. 2021;**13**(21):2951. DOI: 10.3390/w13212951

[20] Li Y, Wang Y, Wang B, Wang Y, Yu W. The response of plant photosynthesis and stomatal conductance to fine particulate matter (PM2.5) based on leaf factors analyzing. Journal of Plant Biology. 2019;**62**:120-128. DOI: doi.org/10.1007/s12374-018-0254-9

[21] Jeanjean AP, Buccolieri R, Eddy J, Monks PS, Leigh RJ. Air quality affected by trees in real street canyons: The case of Marylebone neighbourhood in Central London. Urban Forestry & Urban Greening. 2017;**22**:41-53

[22] Hirsch M, Böddeker H, Albrecht A, Saha S. Drought tolerance differs between urban tree species but is not affected by the intensity of traffic pollution. Trees. 2022;**27**:1-21

[23] Norton BA, Coutts AM, Livesley SJ, Harris RJ, Hunter AM, Williams NS. Planning for cooler cities: A framework to prioritise green infrastructure to mitigate high temperatures in urban landscapes. Landscape and Urban Planning. 2015;**134**:127-138

[24] Zölch T, Maderspacher J, Wamsler C, Pauleit S. Using green infrastructure for urban climate-proofing: An evaluation of heat mitigation measures at the micro-scale. Urban Forestry & Urban Greening. 2016;**20**:305-316

[25] Sanusi R, Livesley SJ. London plane trees (*Platanus x acerifolia*) before, during and after a heatwave: Losing leaves means less cooling benefit. Urban Forestry & Urban Greening. 2020;**54**:126746

[26] Schwaab J, Meier R, Mussetti G, Seneviratne S, Bürg C, Davin EL. The role of urban trees in reducing land surface temperatures in European cities. Nature Communications. 2021;**12**(1):1-11

[27] Perkins SE, Alexander LV. On the measurement of heat waves. Journal of Climate. 2013;**26**(13):4500-4517

[28] Watts N, Amann M, Arnell N, Ayeb-Karlsson S, Belesova K, Boykoff M, et al. The 2019 report of the lancet countdown on health and climate change: Ensuring that the health of a child born today is not defined by a changing climate. The Lancet. 2019;**394**:1836-1878

[29] Sallnow L, Smith R, Ahmedzai SH, Bhadelia A, Chamberlain C, Cong Y, et al. Report of the Lancet Commission on the value of death: Bringing death back into life. The Lancet. 2022;**399**(10327):837-884

[30] Krutilla JV. Conservation reconsidered. The American Economic Review. 1967;**57**:777-789

[31] Baveye PC, Baveye J, Gowdy J. Monetary valuation of ecosystem services: It matters to get the timeline right. Ecological Economics. 2013;**95**:231-235

[32] Bunse L, Rendon O, Luque S. What can deliberative approaches bring to the monetary valuation of ecosystem services? A literature review. Ecosystem Services. 2015;**14**:88-97

[33] Ehrlich P, Ehrlich A. Extinction: The Causes and Consequences of the

Disappearance of Species. New York: Random House; 1981

[34] Ibes DC. Integrating ecosystem services into Urban Park Planning & Design. Cities and the Environment. 2016;**9**(1):Article 1

[35] Delgado LE, Marín VH. Ecosystem services: Where on earth? Ecosystem Services. 2015;**14**:24-26

[36] Castanho RA, Cabezas J, Naranjo JM, Gallardo JM, Fernández-Pozo L, Genç SY, et al. Assessing Ecosystem Services Delivered by Public Green Spaces in Major European Cities. Landscape Architecture-Processes and Practices Towards Sustainable Development. London: IntechOpen; 2020

[37] Lazarus E, Zokai G, Borucke M, Iha K, Morales JC, Wackernagel M, Galli A, Gupta, N. Working Guidebook to the National Footprint Accounts. Global Footprint Network. Rep. 127. 2014

[38] Lippolis L. Vaggio al Termine Della Cità – La Metropoli e le Arti Nell'Áutumno PostModerno (1972-2001). Genova, Italia: Elèuthera; 2016

[39] McDonnell MJ, Pickett STA. Humans as Components of Ecosystems: Subtle Human Effects and the Ecology of Populated Areas. New York: Springer-Verlag; 1993

[40] Pickett STA, Cadenasso ML, Grove JM. Resilient cities: Meaning, models and metaphor for integrating the ecological, socio- economic and planning realms. Landscape and Urban Planning. 2004;**2004**(**69**):369-384

[41] Nowak DJ, Greenfield EJ. Declining urban and community tree cover in the United States. Urban Forestry & Urban Greening. 2018;**1**(32):32-55. DOI: 10.1016/j.ufug.2018.03.006

Chapter 2

Urban Green Spaces Prospects and Retrospect's

Akanksha Sangwan, Anjali Saraswat, Nand Kumar, Satish Pipralia and Ashwani Kumar

Abstract

Urban green spaces (UGSs) are integral structural elements of the city's existence and are promoted to serve diverse functions such as recreation, ecological services, regulation of ambient air temperature and hydrological regime pollution abatement, social inclusion, enhancing amenity value, etc., in sum, they help minimize the adverse impacts of urbanization on the environment and improve citizens' habitable experiences. However, due to competing economic interests and demand on land for various purposes such as residential, commercial, industrial and institutional, UGS in cities take a back seat and are seldom given desired attention, and Indian cities are no exception. The challenges faced by UGS are many and include Land availability, quantity, quality, distribution, accessibility, lack of intended purpose and stakeholder participation. Therefore, it is imperative to explore strategies for planning UGS to ensure their adequate provision and maintenance in cities. Unlike Western countries cities in India have diverse classification and hierarchy of green spaces, therefore, this text examines typology of green spaces, their provisions, associated benefits, emergent concerns in the Indian context through case studies and concludes with plausible strategies for inclusion in urban planning framework.

Keywords: urban green spaces, open and green space, urban planning, urban green

1. Introduction

"Greenspace" the word itself is so vast for the academicians that it has been used by researchers differently in different contexts. With the revolutions in the research of sustainability, the term "green" not only symbolizes trees and vegetation but is also used as an adjective for environment-friendly built environments and even technologies. In urban planning, the term "greenspace" is referred to as the vegetation cover of the spatial area. Greenspace is an urban space that fulfills various esthetics, air purification, conservation of ecology, etc. Urban green spaces (UGSs) in cities exist as natural or semi-natural, managed parks and gardens, supplemented by scattered vegetation pockets associated with roads and random green spaces [1].

Some traditional and cultural parameters, including health, ecological, social, and recreational, remain a major deterministic factor for the optimum usability of these spaces. These UGS can be accessible or inaccessible within city areas. Accessibility of green spaces is an important aspect to assess its impact. There are many UGS within the city area which are inaccessible to the public, and thus citizens are not able to avail themselves of its best benefits. In such cases, green space may not perform its function. Therefore, it is necessary to understand the context of green spaces in urban planning. To further elaborate the green space, 'public open spaces' is another term that can further help understand the concept, functionality, and psychological impact of the population affected by green spaces. UGS include natural elements like green belts, soil, water, parks, etc. have a positive effect on human wellbeing, as shown in **Figure 1**.

The sustenance of human life on Earth is entirely due to a natural environment and large biodiversity functions. Fresh air, water, fruits, woodlands and minerals are all provisioned by various ecosystem services. Green spaces in cities serve as a natural gathering place for the community, fostering social interaction and integration. They also promote individuality and belongingness to urban areas. The effectiveness of UGS to attenuate air pollution is enhanced by vegetation density. Urban greenery provides a safe and healthy atmosphere for walking, jogging, and running and a conducive environment for social contact and physical and leisure activities. Consuming the maximum benefits of city greens requires them to remain unaltered by the urban infrastructure such as the buildings, highways, and other infrastructural components. The current times require greens to be planned as ecological functional spaces coexisting to support the human functions of recreation, esthetics, leisure activities and conserving environmental values.

Vacant built-up space remains as elapsed wasteland or gaps between buildings and other constructions. These spaces have a high potential for reconstruction and repurposing by integrating them into the public, and for creating stunning spaces by distinguishing their specific character. For example, the exact characteristic of a former railway track is that it connects two regions and can be renovated into a green corridor. Depending on their location, abandoned areas can be converted into different facilities. Every city has such vacant spaces that are waiting to be adapted to the current urban fabric so that they can be part of the total cityscape.

An environmental justice issue that affects so many communities is the lack of green space. Most cities have green areas throughout them for the health of the citizens. In areas that are predominantly lower income or a minority group tend to

Figure 1.
UGS in Chandigarh city, India (Chandigarh is the first the planned city of India by architect-planner Le Corbusier and is famous for its green spaces integration in the master plan). (Source: author).

have less green space. Green spaces are very important to maintain good air quality and promote exercise [2]. There are many cases of this environmental justice problem that were acknowledged and even fixed. A few examples are Tartu, Estonia, Faro, Portugal, and Phoenix, Arizona. Green space is something everyone should have equal access to, but due to the environmental justice problem certain groups are deprived of easily accessible green space.

Living near a green space might even help you live longer. A scientific review published in The Lancet Planetary Health found urbanites living near a park or a garden had a lower risk of premature passing. In their work, the researchers used a vegetation index to measure the density of greenery in locales. Using their scale, infertile areas composed of rocks or sand would score closer to a zero, while an area like a lush tropical rainforest would score closer to one.

2. Methodology

The chapter holistically covers and generalizes concepts and concerns associated with UGS across urban planning domain. The chapter is written based on the review and analysis of secondary data available in the form of published literature from reputed data sources, government reports and based personal observation of the author may during the course of study.

3. Benefits of urban green spaces

The green space can be differentiated from open space as it constitutes only a two-dimensional land area that has not been modified to buildings, highways, and other infrastructural components. Greens should be mainly classified into an ecological function space (flora and fauna, physical infrastructure (drainage, stormwater management, and water quality conservation), and human function (provision for recreational, aesthetical, emotional, and leisure activities, preservation of environmental values, and solar access). Green space may consist of a vast range of pre-hold and leasehold land tenure with different use and access rights. Green space may also find its share in further land-use planning.

In the early 1700s, garden squares and parks were firstly introduced in urban areas. In the late eighteenth and early nineteenth century, the parkway movement, headed by Frederick Law Olmsted, led to the belief that green space was an important cultural and social element of the urban form. The idea of integrating green space into the urban development process came out with the regeneration of the environmental movement in the 1960s [3]. Although public access is defined for various green spaces, green space does not imply public accessibility automatically. Here the green space categories the open spaces which support human socio-cultural and health requirements, biodiversity preservation as well as discussion related to historical green space perspective [4, 5].

There are different types of green spaces like food production areas, national parks, grasslands, green belts, and playfield (like a golf course) that are predominately utilized for corps and live stocks, habitation to animals and medicinal herbs, cattle grazing, the barrier between the pavement and human settlement, and sports activity respectively. Green space categories intermingled with the next one, parks and gardens, where the needs of individuality for recreation and enjoyment are fulfilled

[6, 7]. **Table 1** shows the Indian UGS typology according to the designated documents with different contexts, categories, classification, and other hierarchies.

3.1 Categorization of green spaces according to the cities

Many cities in India have a categorization of spaces that are derived from literature (**Table 2**). And then, there is a hierarchy of classification of categorized spaces. The planned cities have systematic cataloging of green spaces like the recreational spaces, green areas and tourist areas. Each town has its USP (unique selling point), and it offers the spaces it has according to its function. **Table 3** show cases the different cities of India and how they have their green spaces categorized, their further classification and the hierarchy.

Figure 2 shows an overall view of UGS and countries' happiness worldwide. This map highlights regional differences in the green space distribution due to climate; countries near the Equator in tropical climates have relatively high UGSs, while countries in the 20–30 latitude range have exceptionally low UGSs due to the dry climate. The UGS increases with latitude in higher-latitude regions. On the other hand, Northern and Western European and North American countries display relatively high happiness scores. Western Asian countries also show relatively high happiness with a low UGS, indicating that the relationship between happiness and green space is not trivial.

Documents	Context	Category	Classification	Hierarchy
UDPFI (1996)	Open spaces	Recreational facilities	Parks and open spaces Sports centers and playgrounds Botanical and zoological parks Water bodies/other natural features Places of tourist interest	Housing cluster Sector Community District Sub-city center
URDPFI (2014)	Open spaces	Recreational space Organized green Other common spaces (such as vacant lands/ open spaces including flood plains, forest cover etc. In plain areas)	Recreational	Recreational P-1 Playgrounds/ stadium/sports complex P-2 parks and gardens—public open spaces P-3 multi-open space (Maidan) Organized green Housing cluster Neighborhood Community District/zone Sub-city center
Urban Green Guideline (2014)	Green spaces	Urban greens	Reserved forest Protected forest District park Neighborhood park Tot-lots Playgrounds Green belts (buffer) Green strip	Tot lot Playground Neighborhood park Community park

Table 1.
Indian UGS typology.

Typology	International typology	Indian typology
Amenity green space	Recreational green space Parks and gardens Outdoor sports areas Green street space Private green space Green roofs Courtyard Greenery at housing estates Greenery at commercial buildings	Reserved forests Protected forests National parks Sanctuary
Functional green space	Green trail Car parking space Pedestrian area Stormwater retention area Old landfills/dumps Productive green space Remnant farmland City farms Cemeteries/churchyards	City parks District parks Neighborhood parks Tot lots
Linear green space and open spaces	River and canal banks Transport corridors—road/rails	Green belts (buffers) Green strips Transport corridors
Natural and semi-natural spaces	Wetlands Woodlands Vacant land Post-industrial land	Playgrounds
Green corridors	Tree belts Canal and riverbanks Disused railways	
Allotments, community gardens, and urban farms	Allotments Community gardens City farms	
Outdoor sports facilities	School playing fields Community playing fields	
Provision for children and young people	Facilities for young people Community parks Children's play area	
Privately usable areas	Facade greening Roof garden Front garden Courtyard Balcony	
Leisure areas	Campsite Multifunctional leisure facility Sports ground	
Eco-sensitive areas	Waterlogged Marshy Swampy	

Table 2.
Categorization of green spaces according to the typology.

City name	Categorization of spaces	Classification of categorized spaces	Hierarchy of classification of categorized spaces
Jaipur	Recreational space/tourism zone	G1 eco sensitive areas G2 green zones G3 parks	Parks Gardens Stadium Sports ground
Jodhpur	Recreational space	Parks and open spaces Stadium and play grounds Fair and tourism Public entertainment	No further classification
Sriganganagar	Recreational space	Parks, open spaces and playground stadium	No further classification
Chandigarh	Recreational space/green areas	Sports facilities Cremation ground Cultural facilities	Organized open space Forest Agriculture
Naya Raipur	Recreational space	Open spaces	Film city Parks and play areas Stadium and sports complex Nature resort/theme park Reserved forest Botanical park, jungle safari
Lucknow	Recreational space/green areas	Parks and open spaces	No classification
Pune	Recreational space	Community hall Museum/theaters Parks and open spaces	No classification
Varanasi	Recreational space	Gardens, open spaces and urban forest	No classification
Panaji	Recreational space	City level parks and playgrounds	Housing area park Neighborhood park Community park District park
Pudducherry	Recreational space/tourism zone	Open space	Parks Gardens Stadium

Table 3.
Categorization of green spaces according to the typology.

The distribution of UGS and happiness over the world. (a) The map of UGS and happiness in 60 developed countries. The size and color of circles represent the level of happiness and UGS in a country, respectively. The markers are placed on the most populated cities of each country. (b)–(d) The histograms of (b) happiness, (c) UGS and (d) logarithmic GDP per capita (log-GDP). We use the logarithm of the total NDVI per capita as an indicator of UGS and the logarithm of GDP per capita as a measure of wealth.

4. The uses of green spaces (according to the place)

The usage of Greenspace for recreation and leisure purposes led to the identification of its economic, cultural, environmental, and social values, which further

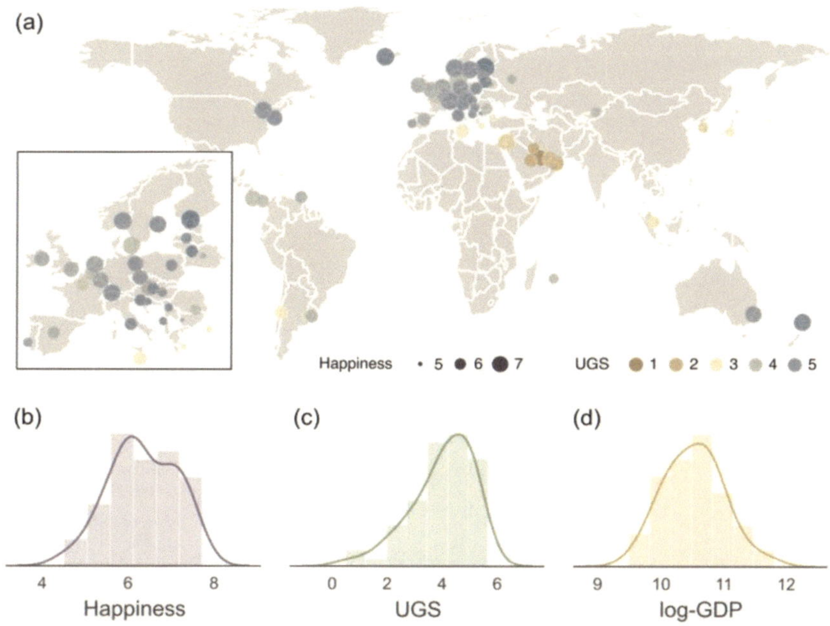

Figure 2.
UGS and happiness index of countries worldwide (a) The map of urban green space and happiness in 60 developed countries. The size and color of circles represent the level of happiness and urban green space in a country, respectively. The markers are placed on the most populated cities of each country. (b) Histogram of happiness (c) Histogram of UGS (d) Histogram of log-GDP.

increased the attention on its management and planning. In this chapter, the focus is on the relationship of Greenspace to human interaction on various levels of livability, recreational purposes and improved quality of life. The other perspective describes urban vegetation, such as parks, yards, and gardens related to a green stuff kind of open space. This perspective may be defined as a subsidiary of a comprehensive notion of Greenspace, i.e., limited to the built-up environment subsection of open space.

UGS are critical for making our cities sustainable and energy-efficient [8]. However, for UGS to contribute to the optimum, they have to be planned, designed, developed and managed/maintained appropriately so that they are accessible both in terms of area and population coverage. It is a fact that urbanization in India will continue unabated. The Urban Greenspaces generate diverse ecosystems of substantial significance for human wellbeing and human activities, shaping their dynamics. Many green spaces in cities disconnected from the wider environment tend to lose biodiversity due to continuous construction activities. Hence, protecting green spaces in isolation will often fail to sustain the capacity of urban ecosystems to generate value and have to be well integrated into the overall city landscape.

Figure 3 shows the distribution, parts and sub-parts of the UGS.

Furthermore, the socio-cultural, functional, health-related aspects are labeled for a specific place, i.e., "Greenspace", as a park or recreational area [9]. Sprawling population, commerce, industries, transportation accelerated the gross domestic production, varied product availability to the consumers, and more excellent connectivity. With the high-rise demand of time, the

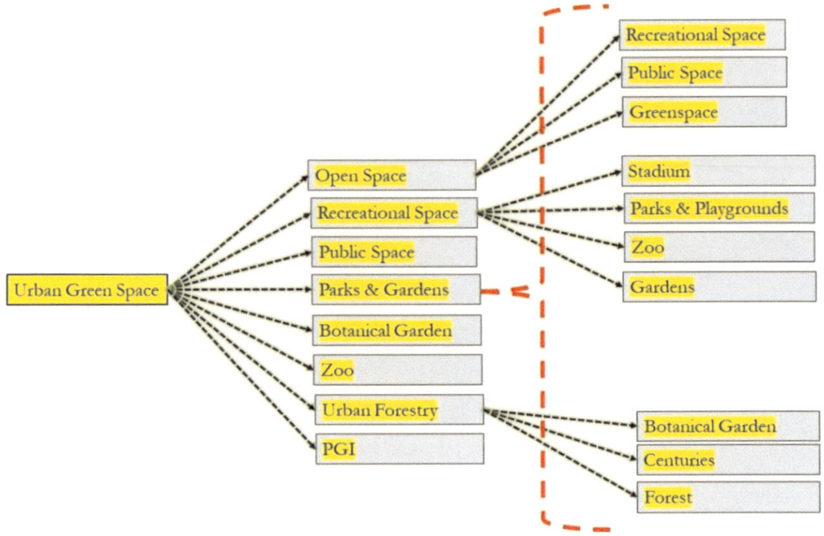

Figure 3.
Synonymous nature of UGS.

researchers explored the correlation between these sectors and related aspects of "Greenspace".

4.1 Regional green space use

Typology	Definition
Natural land	Unoccupied/unused lands, left unturned by the authorities or the locals, come under this category. Grasslands are also a big part of the natural land group.
Green belts	Green belts of buffers include green girdle, park belt, rural belt, agricultural belt, country belt etc., which generally refers to a stretch between towns or regions separating one from the other. These areas are dominantly farmlands as they support agriculture and related functions. Green belts are established to keep in check the growth of the built-up regions, preserve neighboring towns from merging and also maintain a unique character of a town.
National parks	National park is an area that is protected and conserved due to the presence of remarkable natural flora, fauna, geological formations and natural scenic spots.
Reserved forests	Area duly notified under the Indian Forest Act, 1927 or the State Forest Acts having complete protection. All activities inside are prohibited unless expressly permitted.
Protected forests	They are found in urban and peri-urban areas secured by fencing or constructing a compound wall, or both. Here no construction activity is allowed.
District parks	It is a designated term per the hierarchy of green spaces in an urban city. It is a prominent use of a vast green area and developed as a crucial green space in a city. As per UDPFI guidelines, one district park serves 500,000 of the population in plain areas.

4.2 Urban green space use

Typology	Definition
Campsite leisure	Land that is dedicated to camping acquires a significant part of the green space category, especially in the trekking zones of any country

Typology	Definition
Wetlands/ marshy land	A wetland is a part of the ecosystem flooded by water, either permanently or seasonally. Wetlands are considered amongst the most biologically diverse of all ecosystems, serving as home to a wide range of plant and animal species.
Functional green space	Spaces like cemeteries, pedestrian area, landfills, farms, etc., are functional green spaces. These areas acquire land which is usable for the users.
Playing fields	As part of any institution/university/school, fields serve as a decent open area for any city. It enables the users to use it for any leisure/formal purposes.

4.3 Neighborhood green space use

Typology	Definition
Linear green space	Areas like pedestrians, river canals, street trees are included in Linear Green Space. It adds to the esthetic and usability value of any city.
Green stripe	A green strip is developed on vacant land, for instance, land under high tension power supply lines. It is also developed along the arterial roads, separating residential areas from other land uses.
Green roof/ facade	Green roofs are the most popularly deployed form of roofs; they are generally lightweight and low cost. They are purposely fitted or cultivated with vegetation. They are also be known as living roofs, eco-roofs or vegetated roofs. They can reduce both heating and cooling loads in buildings. This has positive implications in terms of their energy consumption. The green roofs and facades increase the esthetic value of the scape.
Neighborhoods parks	It is developed at the neighborhoods level for a population of 10,000. Neighborhoods Parks are conveniently located within the developed residential areas, preferably within walking distance. It is planned on a site of 2000−4000 m^2

5. Case studies

Indian city	Typology
Chandigarh: The capital city has a forest and tree cover of 35.5% in its 114-km^2 area. Chandigarh has become the greenest city in the country. The tree cover saved the city from becoming an all-concrete jungle, a fate that has overcome many Indian towns.	The green belts of Chandigarh also facilitate a healthy population of diverse birds. The parks along the spine of Chandigarh include Rajendra Park, Bougainvillea Garden, Leisure Valley Garden, Zakir Rose Garden, Shanti Kunj, Bamboo Valley Garden, Bulbous Hibiscus Garden, Fragrance Garden and Dahlia Garden.

See **Figure 4.**

Indian city	Typology
Delhi: The capital of India is one of the greenest capitals in the world due to the consistent emphasis on greening and strict monitoring of tree cutting.	Recently, the parks and garden society has been set up to coordinate the greening activities in Delhi. The city has some well-maintained parks and gardens like Lodhi Gardens, Mughal Gardens, Deer Park, Budha Jayanti Smarak Park, Indraprastha Millennium Park and The Garden of Five Senses besides the Ridge.

Figure 4.
Plan of Chandigarh.

See **Figure 5.**

City abroad	Typology
New York – Central Park: Central Park is an urban park in New York City located between Manhattan's Upper West and Upper East Sides. It is the fifth-largest park in the city, covering 843 acres (341 ha). It is the most visited urban Park in the United States, with an estimated 42 million visitors annually as of 2016, and is the most filmed location in the world.	The Park has natural-looking plantings and landforms, having been almost entirely landscaped when built in the 1850s and 1860s. It has eight lakes and ponds created artificially by damming natural seeps and flows. Several wooded sections, lawns, meadows, and minor grassy areas. There are 21 children's playgrounds and 6.1 miles (9.8 km) of drives.

See **Figure 6.**

City abroad	Typology
London – Hyde Park: Hyde Park is famous for being the largest Park in Central Park and the royal parks of London and its speaker's corners.	Hyde Park is a Grade I-listed major park in Central London. It is the largest of four Royal Parks that form a chain from the entrance of Kensington Palace through Kensington Gardens and Hyde Park, via Hyde Park Corner and Green Park, past the main entrance to Buckingham Palace. The Park is divided by the Serpentine and the Long Water lakes.

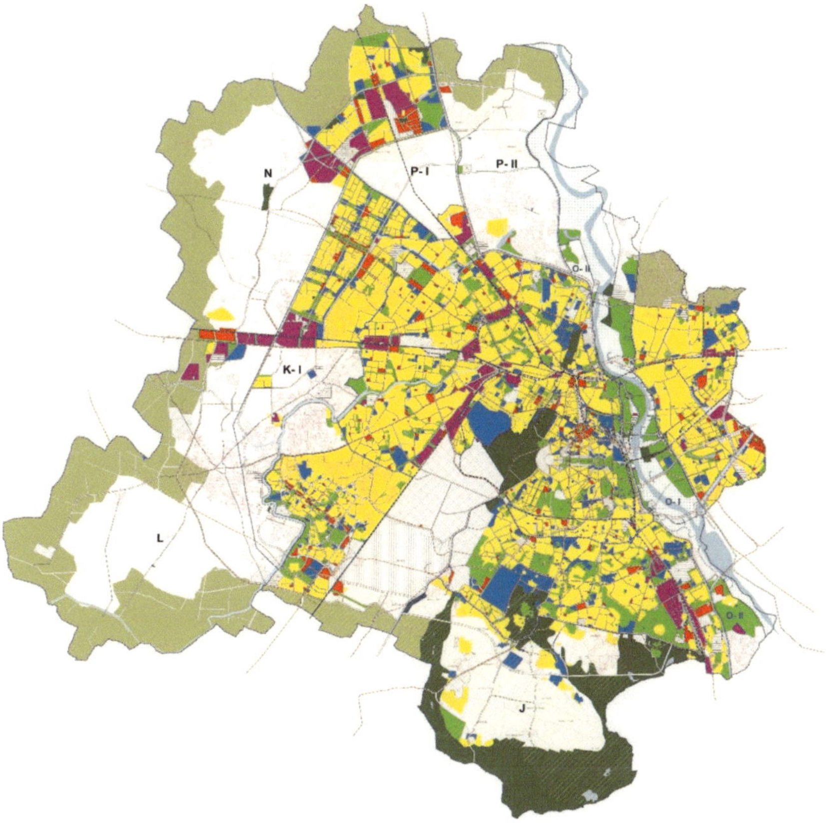

Figure 5.
Plan of Delhi.

Figure 6.
Plan of New York Central Park.

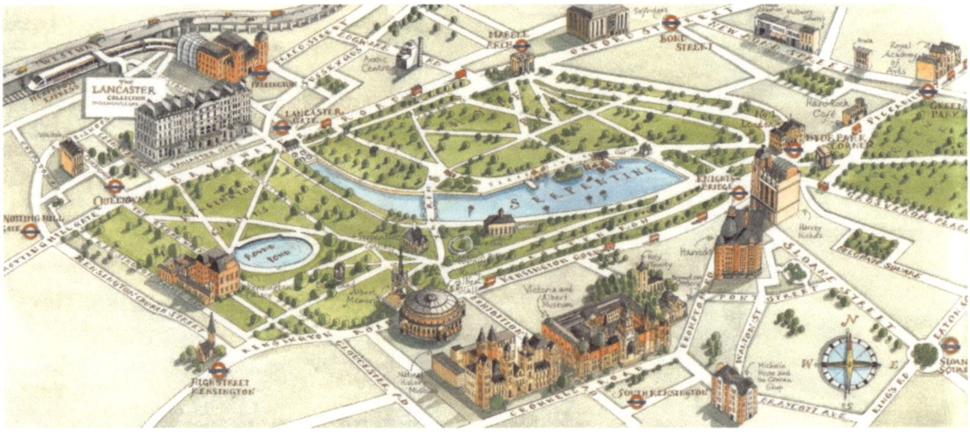

Figure 7.
Plan of London Hyde Park.

See **Figure** 7.

6. Benefits of the green spaces

Different direct accessibility Greenspace typologies provide ecological benefits such as ambient temperature, air and noise pollution reduction, water harvesting, and a barrier between pavement and locality. Green Space provides indirect health benefits instead of direct benefits. These green spaces offer various herbs, vegetables, cereals, fruits, etc., supporting human health. Flourished greenspaces promote tourism, health wellness product vending and fetch high real sale values for the properties near parks and greenways. Apart from these direct benefits, many researchers have established many indirect benefits of Greenspace, which will be discussed later in this chapter.

6.1 Social benefits

Recreation opportunities, Areas that are culturally and historically valuable. It has an impact on physical and mental health. Green spaces provide a refreshing contrast to the harsh shape, color, and texture of buildings and stimulate the senses with their simple color, sound and smell. Particular types of green space may offer a bigger diversity of land uses and opportunities for a wide range of activities, help to foster active lifestyles, and can be of real benefit to health.

6.2 Esthetic/architectural benefits

Green Spaces add to the Landscape features of any cityscape; screened views are formed from different angles of the city. Growing trees and experiencing nature is a path to a positive passive lifestyle.

6.3 Climatic and physical benefits

Air pollution reduction, Sound Control, Glare and Reflection reduction. Impacts on urban climate through temperature and humidity control. Urban forests act as temperature buffers providing shade in the summer and windbreaks in the winter in addition to reducing noise pollution and CO_2 levels and providing a habitat for wildlife [10]. Urban greening improves air, water, and land resources by absorbing air pollutants, increasing water catchment in floodplain surfaces, and stabilizing soils.

6.4 Ecological benefits

Biotopes for flora and fauna in any urban environment is benefitted and, in turn, help the environment. Trees absorb pollutants; moderate the impact of humans, absorbing pollutants and releasing oxygen. They contribute to maintaining a healthy urban environment by providing clean air, water and soil. Green vegetation has been shown to lower wall surface temperatures by 17°C, which led to a reduced air conditioning load by an average of 50%. They improve the urban microclimate and maintain the balance of the city's natural urban environment [11]. They preserve the local natural and cultural heritage by providing habitats for various wildlife and conserving a diversity of urban resources.

6.5 Economic benefits

Tourism is increased due to the aesthetical change in any scape. The property value is increased—the value of market-prices benefits [12]. Property owners value urban greenery by the premium they pay to live in the neighborhood of UGS and public parks. Plots and flats adjoining Park add to the value. In densely populated areas, this effect is even more noticeable. For example, the view of green spaces and proximity to water bodies increases the real estate prices. The impact of neighborhood parks on the transaction price of multi-storied residential units in cities illustrates that neighborhood parks could increase the cost.

7. Value added from green space

7.1 Social value

Humans appear to adapt to almost anything in their environment, including air pollution, noise and environmental dullness. Man's apparent ability to adapt may be his most significant liability. On the one hand, it allows him to adjust to slowly developing adverse conditions; on the other, it could easily threaten his survival. Adaptability permits human adjustment to damaging environmental circumstances we aren't aware of. When environmental conditions deteriorate to a point where they are readily apparent, it may be too late to reverse the cycle. If an individual has a feeling of wellbeing in the built environment that surrounds him, it is reflected in his activities. The type of space that covers us can either stimulate or be in habitus. Some have tried to isolate how spaces (man-made and open space) play a behavior role in our way of life and specific ways of thinking. While these studies can be rationalized from many perspectives, a case can be made that the type and amount of spaces in our urban areas do, to some degree, shape our social behavior. As urban life becomes more stressful, the more significant the influence of open space [13]. Stress is most pronounced in low-income neighborhoods.

7.1.1 Money and race

Higher-income neighborhoods outside our major cities tend to be less densely populated, which may reduce the resident's perceived need for public open space. Often, the lower a neighborhood's economic scale is, the less likely it becomes for that neighborhood to have the available space and recreational activities required for good physical and mental development.

7.2 Economic value

Placing an economic value on urban open space related to public benefits is a difficult task. Land value in a free enterprise system depends on supply and demand. The value of the real estate is directly proportional to the market demand, potential use, and rights of ownership within its geographic locality. Publicly owned open space used for recreation does not fit into the regular private land market. The parcel's value as open space cannot be measured in dollars in the same way as land or improvements for a commercial venture.

7.2.1 Fees and rents

One method of estimating the economic value of public open space is to monitor the success of a program and determine the revenue that is directly or indirectly generated by it [14]. At the same time, public parks are primarily supported by tax dollars appropriated through the general fund.

7.2.2 Value of nearby property

More indirectly, open space generates revenues by enhancing the value of adjacent private properties. For example, a 1977 study in Chicago concluded that property prices were $1000 higher for parcels within one block of an urban park as compared to a similar area. Hammer [15] estimated that for each acre of a public park adjacent to a stream, sur-rounding private property values would increase an average of $2600.

7.3 Ecological value

Ecological considerations directly related to land and resource preservation often are neglected. These include the function of open space for water-shed management, environmental quality, and esthetic appreciation.

Greenbelts adjacent to natural water bodies can reduce sedimentation, increase preservation, reduce the need for excessive flood control projects and lessen flood damage [16]. Areas of open space can:

- Ensure groundwater recharge.
- Improve water quality by reducing soil erosion and sedimentation.
- Increase recreational benefits, such as nature study, fishing and boating.
- Sustain wetlands, which provide wildlife habitats.

- Absorb peak water discharge and naturally filter some of the suspended pollutants.
- Keep lake and stream levels more constant over the entire year.
- Enhance the community appearance.

8. Urban green space planning and management issues

The benefits of green spaces for our cities are well documented, but they are seldom given desired attention with other computing land uses such as the residential, commercial and industrial use within the urban planning process. Negligence by planning authorities and decision-makers results in not meeting the quantitative and qualitative standards of green spaces and results in their unequal distribution on the spatial scale. Inaccessible green spaces within urban centers also raise concerns regarding city dwellers' utilization. Primarily new developments in urban centers diminish the future opportunities for recreational provision, and most often, the existing green spaces are not very well managed. Various UGS planning and management issues are thus identified in the subsequent subsections.

8.1 Degenerating quality

The quality of green space is a critical determinant to assess its value to the urban population. In generic terms, the quality of green space is understood in terms of its esthetic outlook and the level of facilities served by it. But in the spatial planning domain, the density of vegetation and the spread of tree crowns can be realized as they essentially lead to the several ecological and social benefits associated with green space. Improved physical and mental health, reduced stress levels, and increased happiness and satisfaction amongst citizens indicate the existence of superior quality greens in the city. Scenarios of degenerating UGS quality are well reported across many cities [17]. The city of Bengaluru in India has seen a decrease of 78% in green cover over the course of four decades, along with its deteriorating quality. These scenarios can be related to rising accounts of air pollution, obesity, shrinking health, and social and psychological collapse amongst the citizens. Overall degenerating quality of the greens and cities can be associated with multiple factors such as lack of maintenance, not choosing native trees, urban sprawl on green belts and ecological areas, deterioration of green riverfronts and loss of biodiversity in forests.

8.2 Inappropriate policies and standards

Policy frameworks in an urban planning setup are intended to guide associated decision-makers in the field towards achieving suitable outcomes related to various city infrastructures. Provision, maintenance and broadening of appropriate green covers in cities require policy measures supporting the existence and promotion of several ecosystem services delivered by UGS. Recommendations made by World Health Organization (WHO) suggests the availability of a minimum of 9 m^2 of green space per person, ceases the intent at large of appropriate green space provision policy as it follows the 'one size fits all approach and neglects varying physical and social circumstances across world cities [18]. Furthermore, catering to large densities in urban centres, many cities fail to meet the standard. For example, (see **Table 4**) amongst many Indian cities, only a few cities like Chandigarh, Delhi, and Bangalore meet the WHO standard, and numerous like Mumbai, Ahmedabad, Surat, Chennai, and Hyderabad miss it on a large scale.

The Ministry of Urban Development sets out urban and regional development plan formulation and implementation (URDPFI) guidelines for, Government of India and suggests appropriate urban development standards. It identifies UGS as part of organized greens within the city's social infrastructure and categorizes them based upon their spatial hierarchy as housing area, neighborhood, community, district and sub-city parks (see **Table 5**). The population to be served per unit and associated area requirements are mentioned within these categories. Such standards meet the quantitative aspects of UGS, but attention to their physical distribution and accessibility in regards to providing the serving radius is missed out.

Indian cities broadly vary in topography, and thus while setting out UGS standards and policy measures, it is crucial to incorporate their diverse local climate context and cultural practices, which these standards miss out on largely.

8.3 Inadequate recreational programming

Developing the proper functionality of an UGS is vital to make it enduring. In the context of UGS associating them with the recreational programming during the

City	City area (km^2)	City population in million (census 2011)	Area of UGS (km^2)	Per capita UGS (m^2 per person)
Chandigarh	114	1.05	50	47.37
Delhi	1484	16.34	324.44	19.84
Bangalore	709	8.43	150	17.80
Mumbai	735	18.45	122	6.61
Ahmedabad	464	5.57	21.8	3.91
Surat	326.5	4.46	11.84	3.32
Chennai	176	8.69	11	1.26
Hyderabad	172	7.75	8.72	0.88

Source: author.

Table 4.
Per capita UGS in Indian cities.

S. no.	Category	Population served per unit	Area requirement (ha)
1	Housing area park	5000	0.50
2	Neighborhood park	15,000	1.00
3	Community park	1,00,000	5.00
4	District park	5,00,000	25.00
5	Sub city park	10,00,000	100.00

Source: URDPFI, volume 1.

Table 5.
Categories of UGS in Indian cities as per URDPFI guidelines.

Figure 8.
Inappropriate recreational programming in Al Shalalat park Alexandria, Egypt. (Source: 2003; De Sousa; thinking brownfields into green space in the city of Toronto).

planning process, it drives them to become more functionally sound and exciting and vibrant for citizens use. But a large number of green spaces in urban centers lack on this front as greens existing across the riverfront, green belts and even large cities scale greens are usually found with the least footfall and thus are encroached by slums and squatters [19]. Even in low hierarchy greens such as local or neighborhood parks, the absence of supporting recreational amenities such as sitting area, walking pathways, sports facilities, biking trails, open gyms, etc. are either not found or are not so well maintained. One of the studies in the Alexandria city of Egypt brings forth the lack of recreational programming of UGS and how it negatively affects the usability of city greens by the urban citizens (see **Figure 8**).

9. Strategies for urban green spaces

Improving the UGS scenarios in our cities requires strategic measures with the coordinated efforts of all decision-makers and the stakeholder's participation. These measures would intend to improve the qualitative and quantitative structure of urban greens in our cities, ensuring equitable distribution and making the greens accessible for citizens. Improving the planning process, strengthening the spatial data, enhancing quality greens, applying ecological principles, using altered greening practices and promoting public participation are discussed in the subsections below as the strategies for UGS improvements.

9.1 Improving UGS planning process

UGS assists sustainable development by being environmentally, socially and economically viable. The accustomed green planning process involves using standards approach wherein a set of standards are used while deciding greens provision. The standards approach ensures enough green spaces exist in cities, but it often compromises other vital aspects. UGS should be of satisfactory quality, adequate quantity, well-distributed and variably accessible [19, 20]. The process of determining standards should be appropriate for the greens' of the greens and should be custom made. They are using local intensive methods such as neighborhood index which are more suitable for the green planning process. The planning process needs to be flexible enough to take in the political change occurring in the cities and should integrate a multi-disciplinary approach involving natural, social, economic, planning, and legal components.

9.2 Strengthening the spatial data

UGS are a system of ever-evolving complexity, and to plan them effectively, spatial information is required that is fine, vigorous, and constantly updated. Geographic Information System (GIS) is one such tool that processes the geospatial data from satellite imagery, aerial photography and remote sensors helping to understand the current green spaces needs. It puts together vast amounts of spatial information and thus assists in analyzing the areas of priorities for UGS and determining the feasibility of developing new green sites in our cities [21]. In GIS, one can run numbers on queries, monitor fluctuations, predict environmental effects and thus help urban planners seek amplified visibility into the available data. Improved public participation is also achieved by using GIS tools for planning UGS, as people can get to know the actual ground reality to be achieved.

9.3 Enhancing the quality of urban green space

For adequate UGS planning, the existing green spaces in the city need to be addressed first. The stub semi-natural land with quality vegetation and varied, rich species can produce sound recreational and biodiverse spaces. The choice of vegetation plays a crucial role in determining and enhancing the quality of greens in a space. Choosing a natural species composition that supplements the biodiversity of the area is required. In cities undergoing densification, innovative greening techniques such as green roofs, green walls, street sites, and renaturation are very beneficial to improving green quality [22]. On an architectural scale, green needs to be saved by making effective plans. Also, site trees should not be cut during the construction phase. Proper distribution of UGS within the cities should be ensured to enhance their accessibility and visibility. Efforts to green the city streets should be made using various shrub species in order to improve the overall quality of UGS.

9.4 Applying ecological principles

The efficacious planning of city greens can be achieved by utilizing ecological principles such as greenways, green fingers, and green belts. Using these principles aims at availing the optimal use of UGS geometry, thus enhancing their accessibility and diversity. Greenways act as a UGS management tool and are usually developed linearly across cities, contributing effectively to city greens. They are created along roads, railways, rivers and ridges helping to preserve green spaces and bring green vegetation into urban areas. Green fingers are conceptualized in the shape of human fingers and have green spaces set in a radial form spreading from city centre to periphery [23]. They bring green vegetation into the core of settlements and their adjoining areas, enabling citizens easy access to UGS. Green belts are set out to act as a protective mechanism checking the outward growth of cities. They are usually created in a ring form at city outskirts to prevent urban sprawl. They also acted as city boundaries separating one city from another. They contributed mainly to preserving city vegetation by safeguarding peripheral land for agriculture, forestry and recreation.

9.5 Alter greening practices

Within compact, dense cities, the land availability is less and thus, altering greening practices are required to fulfill the cities UGS requirements. In such cases practice of green roofs and walls comes to the rescue. Green roofs are the roof of a building covered with vegetation [19]. They not just provide esthetically pleasing landscapes but also create a habitat for wildlife and lower air temperature and mitigate the urban heat island effect. Providing green roofs increases property value in addition to many other benefits. Green wall, on the other hand, is a vertical greening typology wherein the walls are covered with vegetation. It can be used to enhance green spaces on land by using effective multi-level greenery designs. To encourage more and more people to use green roofs and walls, various policy measures in terms of economic incentives and tax exemption is required. Also, spreading environmental knowledge regarding them will increase the willingness of more people to implement these alternative greening measures.

9.6 Public participation

The end-users of the UGS are city residents, and thus it is essential to involve them in the planning process. Public participation programs need to be thoughtfully planned and provide all the necessary information to the stakeholders. The urban planners and decision makers need to be open-minded and committed to considering the stakeholder's inputs. To have active participation, the stakeholders need to explain the decision-making process and have a visual presentation clearly. For effective public participation, the demographic structure of the area needs to be identified, and further, the goals for UGS needs to be set accordingly [24]. Developing a questionnaire based on the area profile, demographics and UGS requirements should be formulated. To educate the citizens regarding UGS, workshops, seminars, and expert discussions can be conducted, which will also help the decision makers better identify local needs and demands.

10. Conclusion

With the rapid urbanization occurring in our cities, the need for sustainable development is more than ever. UGS are significant contributors to sustainable development as they provide us with several ecological services. They assist carbon sequestration, reduce the urban heat island effect, act as a barrier for noise pollution and bring down air pollution. The built environment is majorly affected by the UGS as they comprise recreational spaces for citizens that promote social inclusion in communities. Urban planners play a significant role in the provision of UGS in our cities. They are the primary decision-makers guided by the various standards, policies, laws and legislations. But most of the time, UGS seldomly take a back seat in the planning process over other land use such as residential, commercial, industrial and institutional. Therefore, it is essential for urban planning to address the challenges associated with UGS and suggest various strategies and measures. Indian cities face many challenges with the incredible pace of urbanization, such as less quantity of UGS and inappropriate quality. UGS is not equitably distributed in many cities and is not easily accessible [25]. The intended purpose not being served causes dissatisfaction amongst the city residents. To improve the situation of UGS in our cities, it is utterly vital to improve the planning process and at the same time strengthen the spatial data available to the decision-makers [26]. It is primarily important to work on enhancing the quality of UGS and incorporate various ecological principles available to us in the planning process. In cities undergoing densification and reduced land availability, alternative greening practices such as green roofs and walls can be of great help. Above people are all that avail the UGS, so it's imperative to involve them in the planning process. It is equally essential for the youth to take this up in their hands of responsibility, and that is only possible if they have been given a chance, ensured support and most importantly, allotted funds for improving the UGS.

Author details

Akanksha Sangwan*, Anjali Saraswat, Nand Kumar, Satish Pipralia and Ashwani Kumar
Malaviya National Institute of Technology, Jaipur, India

*Address all correspondence to: 2019rar9011@mnit.ac.in

References

[1] Haughton G, Hunter C. Website. Available: Book Reviews: Sustainable cities. By G. Haughton and C. Hunter. Regional Studies Association, Regional Policy and Development Series 7. London: Jessica Kingsley; 1994. viii + 358 pp. £14.95, paper. ISBN 1 85302 234 9. Ecumene. 4(2):243-245. DOI: 10.1177/147447409700400215

[2] Boone C. Cities as Sustainable Ecosystems: Principles and Practices Peter Newman and Isabella Jennings. 2008. Washington DC: Island Press. Paper, $35.00. ISBN: 978-1-59726-188-3. 296 pages. Ecological Restoration. 2008; **1**:274-276. DOI: 10.3368/er.26.3.274

[3] Williamson J. Encyclopedia of Leisure and Outdoor Recreation 200552 Edited by John M. Jenkins and John J. Pilgrim. London: Routledge 2003, ISBN: 0 415 25226 1 £120. Reference Reviews. 2005;**1**: 51-52. DOI: 10.1108/09504120510573918

[4] Konijnendijk CC, Nilsson K, Randrup TB, Schipperijn J. Urban Forests and Trees: A Reference Book. Berlin, Germany: Springer Science & Business Media; 2005

[5] Stanley BW, Stark BL, Johnston KL, Smith ME. Urban open spaces in historical perspective: A transdisciplinary typology and analysis. Urban Geography. 2012;**33**:1089-1117. DOI: 10.2747/0272-3638.33.8.1089

[6] Clunas C. Fruitful Sites: Garden Culture in Ming Dynasty China. Reaktion Books; 2013

[7] Azcorra H, Dickinson F. Culture, Environment and Health in the Yucatan Peninsula: A Human Ecology Perspective. Springer Nature; 2019

[8] Ma D, Xiong H, Zhang F, Gao L, Zhao N, Yang G, et al. China's industrial green total-factor energy efficiency and its influencing factors: A spatial econometric analysis. Environmental Science and Pollution Research International. 2021;**28**:4-5. DOI: 10.1007/s11356-021-17040-1

[9] Taylor L, Hochuli DF. Defining greenspace: Multiple uses across multiple disciplines. Landscape and Urban Planning. 2017;**158**:25-38. DOI: 10.1016/j.landurbplan.2016.09.024

[10] Bonan GB. Forests and climate change: Forcings, feedbacks, and the climate benefits of forests. Science. 2008;**320**:1444-1449

[11] Grunewald K, Li J, Xie G, Kümper-Schlake L. Towards Green Cities: Urban Biodiversity and Ecosystem Services in China and Germany. Springer; 2017

[12] Morano P, Guarini MR, Tajani F, Di Liddo F, Anelli D. Incidence of Different Types of Urban Green Spaces on Property Prices. A Case Study in the Flaminio District of Rome (Italy). Computational Science and Its Applications – ICCSA. Sapienza University of Roma; 2019. pp. 23-34. DOI: 10.1007/978-3-030-24305-0_3

[13] Driver BL, Rosenthal D, Peterson G. Driver BL. Social benefits of urban forests and related green spaces in cites. In: Proceedings of the National Urban Forestry Conference November 13-16, 1978, Washington, D.C. New York, USA: Syracuse; 1980. pp. 98-113

[14] van der Heide CM, Heijman W. The Economic Value of Landscapes. Routledge; 2013

[15] Hammer TR, Coughlm RE, Horn ET. Research Report: The effect of a large park on real estate value. Journal of the

American Institute of Planners. 1974;**40**: 274-277

[16] Elson MJ, Walker S, Macdonald R. The Effectiveness of Green Belt, H.M. Stationery Office; 1993

[17] Shahfahad S, Kumari B, Tayyab M, Hang HT, Khan MF, et al. Assessment of public open spaces (POS) and landscape quality based on per capita POS index in Delhi, India. SN. Applied Sciences. 2019;**1**:368. DOI: 10.1007/s42452-019-0372-0

[18] Brebbia CA, Zubir SS, Hassan AS. Sustainable Development and Planning VIII. WIT Press; 2017

[19] Wotch J, Byrne JA, Newell JP. Urban green space, public health, and environmental justice: The challenge of making cities "just green enough". Landscape and Urban Planning. 2014; **125**:234-244

[20] Haaland C, van der Bosch CK. Challenges and strategies for urban green-space planning in cities undergoing densification: A review. Urban For Urban Greening. 2015;**14**: 760-771

[21] Li X, Ma X, Hu Z, Li S. Investigation of urban green space equity at the city level and relevant strategies for improving the provisioning in China. Land Use Policy. 2021;**101**:105144

[22] Hillsdon M, Panter J, Foster C, Jones A. The relationship between access and quality of urban green space with population physical activity. Public Health. 2006;**120**:1127-1132

[23] Jennings V, Baptiste AK, Osborne Jelks NT, Skeete R. Urban green space and the pursuit of health equity in parts of the United States. International Journal of Environmental Research and Public Health. 2017;**14**:14,1432. DOI: 10.3390/ijerph14111432

[24] Rall E, Hansen R, Pauleit S. The added value of public participation GIS (PPGIS) for urban green infrastructure planning. Urban Forestry & Urban Greening. 2019;**66**:264-274. DOI: 10.1016/j.ufug.2018.06.016

[25] Lahoti S, Lahoti A, Saito O. Benchmark assessment of recreational public urban green space provisions: A case of typical urbanizing Indian City, Nagpur. Urban Forestry & Urban Greening. 2019;**44**:126424. DOI: 10.1016/j.ufug.2019.126424

[26] Bolund P, Hunhammar S. Ecosystem services in urban areas. Ecological Economics. 1999;**29**:293-301

Chapter 3

Green Space Association with Mental Health and Cognitive Development

Prashanti Rao, Somaina Islary and Kapil Natawadkar

Abstract

Increasing urbanization has resulted in urban stress, which not only has affected the cognitive development of children in schools but also adults at workplace. Various research studies have been conducted in this field, and many computational tests were showcased to prove the facts. Still there is a huge gap while designing the school campus and workplace. This study is based on comprehensive portrayal of greenness at both these spaces. The findings of the study shall provide an insight to policymakers and architects with evidence for feasible and attainable besieged interventions such as improving green spaces not only outdoors but also integrating them with inside spaces at school campus and at workplace.

Keywords: cognitive development, green spaces, school campus and urban stress at workplace

1. Introduction

Nature is the eternal source of inspiration for mankind. Even the slightest interaction with the environment can refresh the senses and rejuvenate the mind [1]. Today's societies are increasingly compromising and creating a deteriorating natural environment, with far fewer opportunities to experience satisfying contact with nature as an integral part of our daily lives. The development of urbanization leads to urban stress, which affects not only school children but also adult cognitive development in the workplace.

1.1 Relevance and need of the study

- Facts and concerns related to adult at workplace: As office environments become increasingly stressful and people spend more time at work, short breaks in nature can spur them to be more productive and improve their health overall. The benefits of nature are believed to have a huge impact on adults in the workplace who are exposed to high levels of stress. In the workplace, being close to nature not only improves employee satisfaction, but even promotes an individual's productivity, creativity, and happiness. With more and more employers realizing the importance of employee health, the provision of green space in office buildings is increasingly becoming an important design element. Numerous studies have found that students perceive outdoor spaces on campus as landscaped, well-maintained, and safe; however, they feel that outdoor furniture, sports equipment and lighting are completely inadequate, more importantly; outdoor spaces lack the facilities that can support learning and social connection.

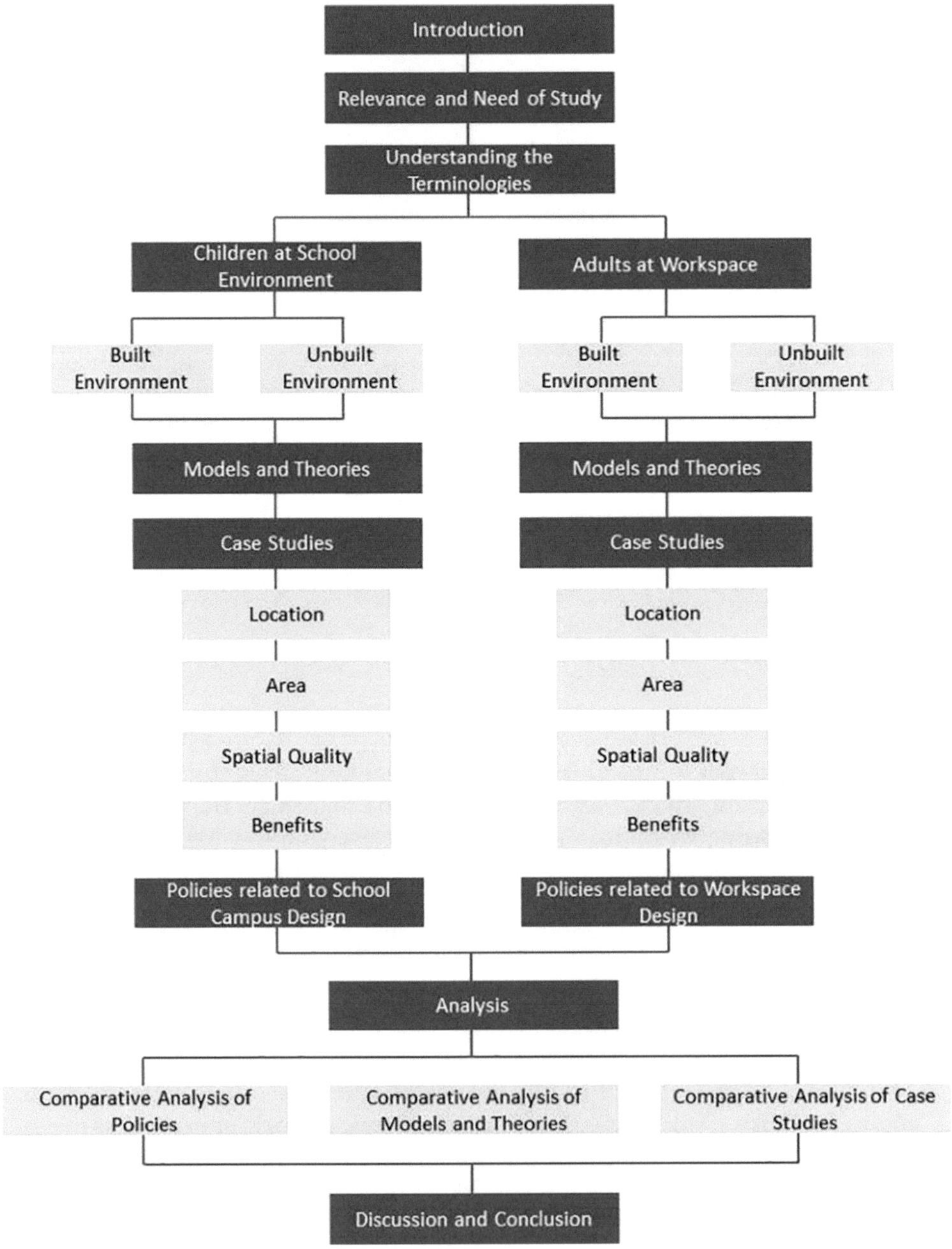

Figure 1.
Methodology (source: Author).

- Facts and concerns related to children at school environment: The educational environment in a campus setting holistically entails the interface of indoor and outdoor spaces. At any level of education, the outdoor space is an essential consideration for designers or school administrators. Similarly, the campus environment can be classified into the physical environment and the social environment even as the duo has significant implications for planning and administration. While the physical environment serves as the physical location where campus life or activities takes place, the social environment is the location for interaction, social norms, and connection among students, staff, and other members of the university community. However, the available shreds of evidence are in agreement with the urban studies, which depict improved performance of the students in classrooms with greener views. It also establishes that students can cope with stressful events when they have proximity to nature [2].

- How and why green space promotes mental health? A philosophical point here is whether it matters how and why green spaces can benefit mental health, or only that it does. The green space and health research area should consider directing its efforts along with these two complementary agendas. It is clear that engagement with green spaces offers benefits in terms of mental health and well-being, and thus green space can act as a preventative measure to promote elite mental health. Therefore, a research program needs to investigate the potential benefits of green space through a range of measures of interest unexplored to date and to examine the extent of these benefits (**Figure 1**) [3].

2. Methodology

2.1 Understanding the terminologies

For humans, nature is an infinite source of inspiration. Even a brief connection with nature can refresh your senses and spirit. Today's culture has resulted in an increasingly degraded natural environment with far fewer opportunities to enjoy pleasant communion with nature in everyday life. The increase in urbanization leads to urban stress, which affects not only the cognitive development of children at school, but also the cognitive growth of adults at work.

- Cognitive development: Cognitive development is the construction of thought processes, including remembering, problem-solving, and decision-making, from childhood through adolescence to adulthood [4]. Through the interaction of inherited and learned elements, cognitive development refers to how a person observes, thinks about, and understands his or her surroundings. Information processing, intelligence, reasoning, literacy skills, and memory are all components of cognitive development.

- Mental health and green space: Green space is a broad term that refers to both controlled and unprotected natural areas, such as protected areas, wilderness areas, and city parks. Green spaces are often created for recreational or esthetic purposes, especially in urban areas. Although global urbanization has reduced access to and participation in green spaces, there is strong evidence that neighborhood green levels are associated with mental health and well-being.

Compared with cities with less green space, people in urban areas with more green space experience less mental discomfort, anxiety, and depression, feel better, and have a healthier cortisol profile.

- Biophilia: Biophilia is defined as a natural human desire to be associated with nature. Natural materials, natural light, plants, nature views, and other natural world experiences are incorporated into the modern built environment through biophilic design, which is an extension of biophilia. Six principles of biophilic design:

    1. Environmental characteristics—Color, water, air, sunlight, plants, animals, and natural materials are all well-known features of the natural world that can be brought into the built environment. Geology and landscapes.

    2. Natural forms and shapes—Biomorphic art, architecture, and design are all examples of natural feature simulation.

    3. Natural processes and patterns—Using time, change, and transitions to transform the sensory experience of a location; complementary contrasts, the play between balance and tension; rhythm, ratios, and scale.

    4. The use of light and space—Understanding how and why people react to light in all of its forms (warm, cool, shaped, filtered, diffused, inside vs. outside) can help you make better decisions about how to use it.

    5. Place-based relationships—The historical, cultural, geographic, spiritual, or ecological value of a location is linked to its meaning.

    6. Evolved human-nature relationships—We have been transformed by our complex relationship with Nature, and we still react strongly to the echoes of our long history [5].

- The therapeutic application of green space for mental health: Green spaces are often used to provide structured therapeutic interventions to vulnerable groups such as at-risk youth, people with dementia or mental illness, probationers, and stressed employees. For example, wild green spaces are used to care for people with dementia and to improve social contact and a sense of belonging. Reducing stress, agitation, anger, apathy, and depression also improves emotional well-being [6].

3. Models and theories

3.1 Piaget's theory of cognitive development

The theory of cognitive development developed by French scientist Jean Piaget (1896–1980) is the most well-known and influential theory of cognitive development. He analyzed that a child's knowledge as being made up of schemas, which are defined as "fundamental units of knowledge used to organize past experiences and provide a foundation for understanding future ones." Through his studies, Piaget declared that cognitive development occurred in four stages throughout one's childhood where (as shown in **Figure 2**),

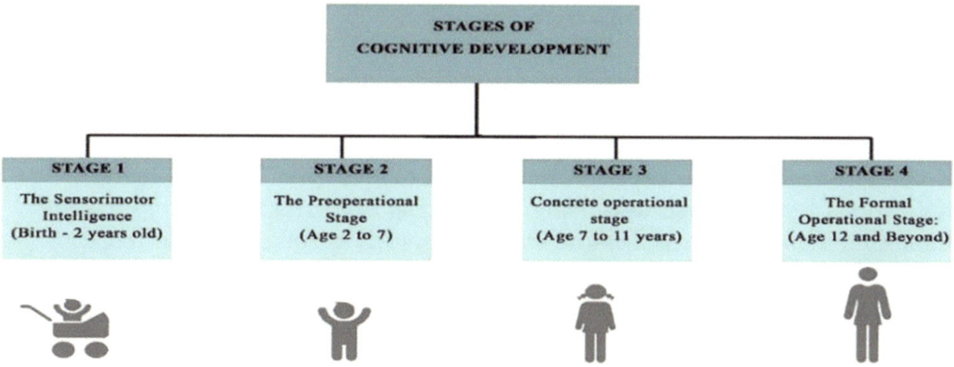

Figure 2.
The four cognitive development stages as per Piaget's theory of cognitive development [7].

- Stages occur in order.
- Children did not skip stages but passed through each one.
- There are visible changes from one stage to the next.
- The stages occur as building blocks, each one using pieces from the last stage.

These stages always actually happen in the same order, and each one builds on the previous one. These four stages are as follows [8]:

- Sensorimotor stage (infancy): This phase acknowledges six substages where Intelligence is exhibited through motor activity without the use of symbols. Around the age of 7 months, children learn object permanence (memory). Physical development (mobility) enables an infant to begin developing new psychological capabilities. At the end of this period, some symbolic (language) abilities are acquired.
- Preoperational stage (toddlerhood and early childhood): Intelligence is demonstrated through the use of symbols, language matures, memory and imagination develop, but thinking is done in a nonlogical, nonreversible fashion throughout this stage, which includes two substages. The dominant mode of thought is egocentric.
- Concrete operational stage (elementary and early adolescence): This stage is characterized by seven types of conservation (number, length, liquid, mass, weight, area, and volume). Intelligence is demonstrated through logical and systematic manipulation of symbols related to concrete objects. Operational thinking develops (mental actions that are reversible). Egocentric thoughts get diminished.
- Formal operational stage (adolescence and adulthood): In this stage, intelligence is demonstrated through the logical use of symbols related to abstract concepts. Early in the period there is a return to egocentric thought.

3.2 Healthy academic green space framework

Embracing therapeutic landscapes as part of academic green space—The Healthy Academic Greenery Framework (HAGF) summarizes the most important relationship between academic greenery and the physical, mental, and social well-being of students.

This allows for a more detailed discussion of the results of using Hofgarten, its implications and perceived well-being. Comparing health experiences with well-established theories of greenery and well-being, such as Völker and Kiestemann's four-dimensional appropriation and attentive recovery theory that extend the concept of the therapeutic environment, compares the results of the Hofgarten study with those of previous studies conducted in different scenarios or in different populations context. The allocation size is no different (as shown in **Figure 3**). It is important

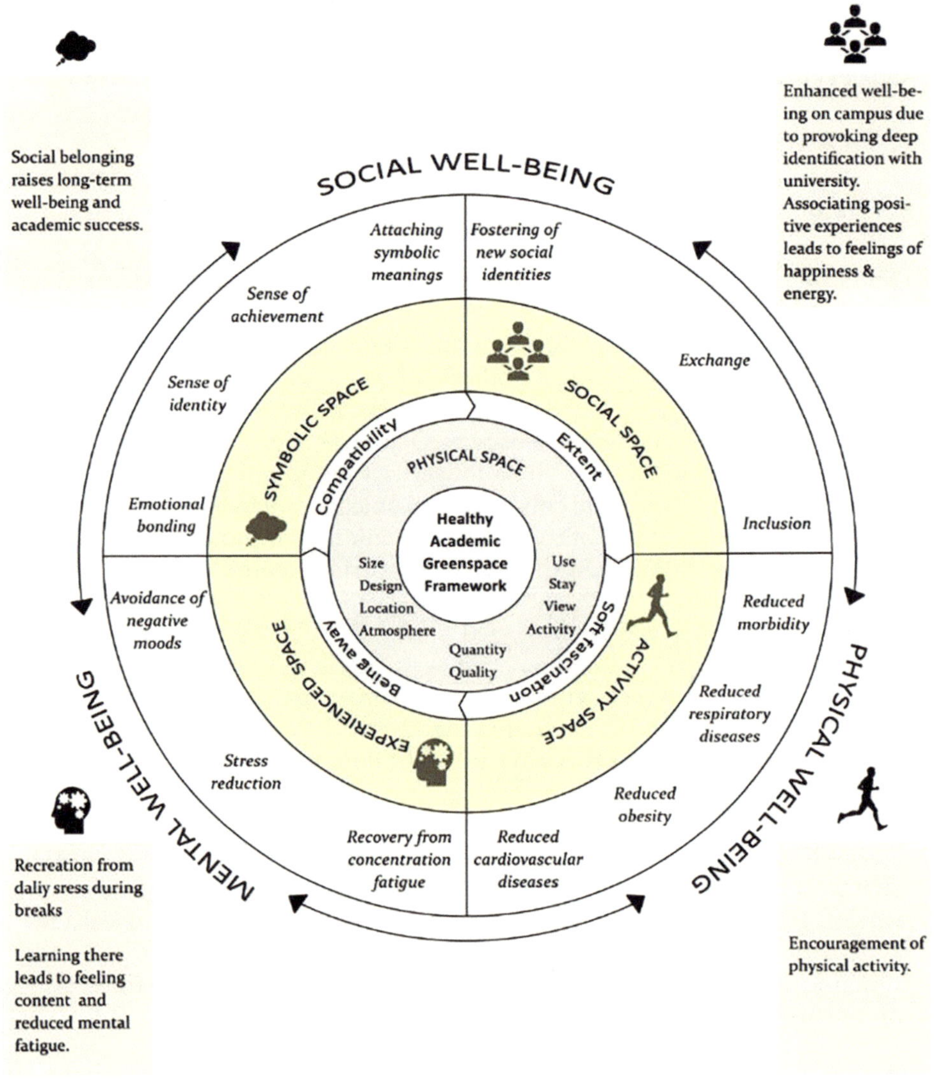

Figure 3.
Healthy academic green space framework (HAGF).

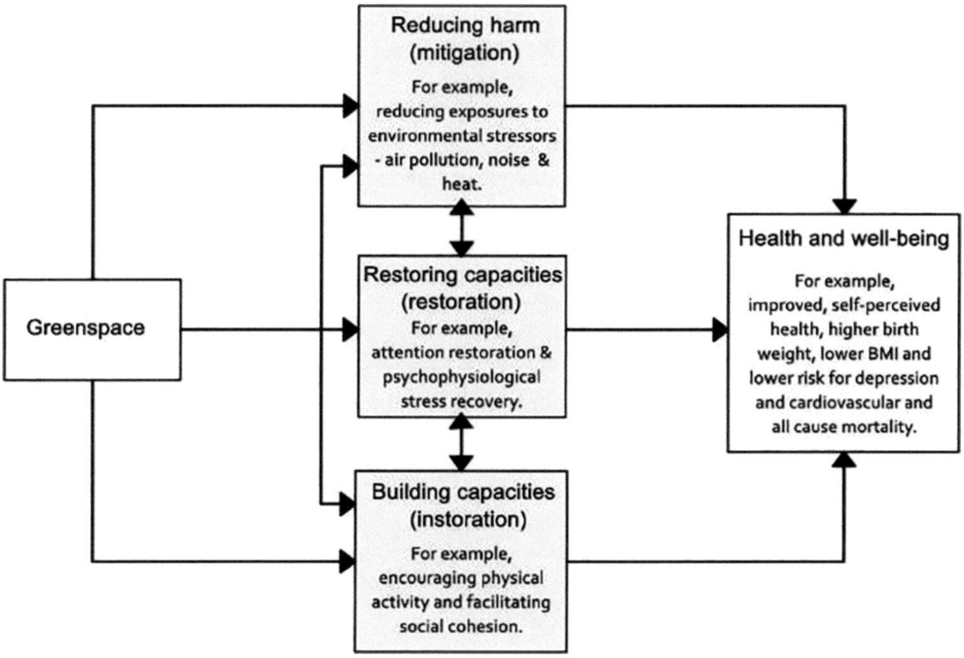

Figure 4.
Three domains of pathways linking green space to positive health outcomes [10].

to recognize that the relevant dimensions are dynamic and intertwined aspects of the interdependent structure of health promotion rather than easily identifiable dimensions. From intangible experiential or symbolic meanings to tactile interactions such as social contacts and physical interactions, these parameters suggest different approaches to health policy [9].

3.3 Potential pathways linking exposure to green space and health outcomes

Green space's health benefits can be explained by a multitude of biopsychosocial pathways. It is possible to divide them into three domains, each focusing on a different function of green spaces (as shown in **Figure 4**).

4. Benefits of green spaces

4.1 Benefits for children

Recent studies have shown that green spaces have positive effects on children's overall well-being, such as mental health, physical fitness, and social skills.

- Physical development: In medicine, childhood is considered an important age for the development of large and fine motor skills in children, and children who play in their natural environment perform better on relevant tests than on regular playgrounds. Childhood overweight/obesity and sedentary lifestyles are common during epidemics. However, exposure to green space has shown a relatively low prevalence of the aforementioned traits. Myopia is less likely to develop from exposure to sunlight and spending time on green school grounds. Research shows that children of all ages enjoy physical activities such as walking and biking when they have more access to outdoor spaces.

- Mental development: A study of the link between green spaces and cognitive development in children found that minor associations, such as having a greener window view, could positively affect attention span and academic achievement, as well as improve imagination, creativity, and intelligence. Notable benefits are: (a) stress relief, (b) improved concentration, (c) increased self-confidence, (d) sense of community and sense of community.

- Social development: Observations have shown that children performed better on attention and spatial working memory tasks after walking in nature than walking on city street [11]. Similar studies have found higher standardized test scores. Socialization and community activities among peers, neighbors, and children are usually done in local parks, which can reduce parental stress and improve children's behavior [12].

4.2 Benefits in workplace

According to a study by the University of Twente, increases in the level of focus and higher productivity are noticed in employees working in offices with green spaces.

- Less sick days: Green spaces carry an effect on health and well-being. A Harvard study revealed that green offices experience 30% lesser sick days when compared with other employees, it was also observed that there were 30% fewer symptoms of sick building syndrome, watering eyes, headaches, dizziness, nausea, heart palpitations, chronic fatigue, tremors, cancer, etc.

- Better office cooperation: Green spaces stand as areas of social interactions. The same goes for office spaces as well. A green office reportedly reduces hierarchical friction, encourages collaboration, cultivates employees' engagement with others, and allows more worker/user relationship-building opportunities.

- More focus: Like children, greenery affects cognition as much as 26% in adults also, when compared with non-greened buildings. Workplaces end up with less productivity when an employee is physically present and mentally absent. In an experimental study, analysis of participants reveals an increase in subjective concentration levels in the case of green office, whereas, in the typical office, no difference was observed over time.

- Energy savings and the environment: The experimental study reveals that participants in the green office perceived an increase in air quality, which was a persistent effect in the long term since plants' respiration naturally contributes to the air quality, temperature, and humidity levels in any area, resulting in power savings.

- Curb appeal: Benefits of green offices remain the same for office or employee turnover, such as attracting people to take up the job or enter a rental agreement [13].

5. Case studies

5.1 Gyaan Centre, Jaisalmer, India

Project Gyaan Centre (Rajkumari Ratnavati Girl's School), 2021
Area 500 m^2
Architect Diana Kellogg

5.1.1 Introduction

The Gyaan Centre is platform to invite renowned women artists, designers, and education advocates to create artwork, host events, and exhibit art installations. Comfortable environment assimilates and provides opportunities to learn to read, write, and develop traditional artisan skills unique to the region. The school is well equipped with the all-necessary infrastructure include classrooms, a library, a computer center, and a bus facility to transport girls from neighboring villages. The school is essential and core structure of Gyaan Centre, an exhibition space and women cooperative revolve around. The Gyaan Centre is symbol of the infinite possibilities of women's strength reminiscing the historic, enchanting and sustainable architecture of the Rajasthan (**Figure 5**).

Figure 5.
Gyaan Centre, Jaisalmer.

5.1.2 Concept

The Gyaan Centre is inspired by feminine symbols across cultures, specifically symbols of strength, encompassing on a structure of three ovals to represent the power of femininity and infinity. The spectacular ellipse structure blends seamlessly into the planes of sand dunes within the region of Jaisalmer, with putting arciform walls, which imitate the organic design of Rajasthan's fort.

5.1.3 Spatial quality

Organization of spaces: Siting and orientation of the ellipse axis with the solar climatic angles, which forms larger and wider shadows through the day to enjoy the central court for various activities (as shown in **Figure 6**). Introvert spaces

Figure 6.
Longitudinal and latitudinal sections to understand the shadow.

considering the climatic consideration courtyard planning inspired from traditional architecture of the region. The height-to-width ratio is maintained by enhancing the high parapet wall; further it imitates the skyline of the fortress in the region.

Use of landscape elements: Large hardscape paved plaza to accommodate functional aspects of school activities, such as assembly, breakout space in between classes, and outdoor learning spaces as well it traps and captures the rainwater runoff inspired from "Tanka," the traditional rainwater harvesting system in the arid climatic region. The relatively small softscape pockets compliment as a visual retreat in the monochromatic floors of courtyards. Introduction of flexible shade structure to improve the usability at time of mid-day hours. Water-sensitive landscape design approach considering minimal and locally sourced vegetation pallet associated with cultural roots of the people.

5.1.4 Indoor outdoor experience

The organic form of the building inspired from the sand dunes, cladded with locally sourced material, and inspired from the traditional building design principles the built mass amalgams with the context. The seamless integration of the indoor and outdoor environment is preserved by slight level difference between courtyards and classrooms access passage (as shown in **Figure** 7). The high ceiling permits thermal comfort by reducing the indoor temperature, the clearstory windows serve to release the hot air and opportunities for filtered indirect light.

5.2 Delta-Plantsoen—urban regeneration and landscape design, Netherlands

Project Christian Elementary Jan van Nassau School (SCOH), 2014
Area 8100 m^2
Architect Daryl Mulvihill

5.2.1 Introduction

Delta-plantsoen considers for redevelopment to become the first playground in the Grevelingenveld, which is a neighborhood square in Rivierenbuurt. The design

Figure 7.
Indoor outdoor experience, Gyaan Centre.

DOI: http://dx.doi.org/10.5772/intechopen.103667

program was challenging considering the unique requirement of the context, however, strategically resolved during the design process by introducing significant landscape elements and experiences. The promising interface and continuous user pool requirements and the tractability in land use allocation where green spaces are allowed to be used as school play yard provide an endless opportunities for communities, cultures, and livable cities.

5.2.2 Concept

The interplay of contrasting experiences, passive naturalist and active urban characters are assembling with the undulating threshold between the two extremes known as "ribbon."

5.2.3 Spatial quality

Organization of spaces: The square is divided in the three broader segments; the outer edge is defined by site extent. The ribbon itself encompasses an intermediate zone as well as encasement to innermost core (as shown in **Figure 8**).

Use of landscape elements: The ribbon, a unique landscape element, is introduced, which acts as buffer between experimental and fundamental play opportunities. The ribbon is confining element to define the extent of urban oasis and provide a contemplative visual retreat for kids and also provide the opportunities for parents to sneak-peak into the wild. The ribbon also hosts a distinctive design program for all the age groups from passive grass mounds around inner core and rock climbing, skating, and leisure seating toward the outer interface facing the active zone. The central natural playground is a space where children are free to construct and destruct their own play spaces from natural materials and fast-growing plants such as willow and reeds.

Figure 8.
Site plan, Christian elementary Jan van Nassau school.

5.2.4 Indoor outdoor experience

Strategically designed access points invite users all cross immediate and larger context. Well integrated with the school interface and allows all-time vehicular conflict free access for all age kids. The interwoven circulation allows the thoroughfare and facilities natural surveillance on the user (as shown in **Figure 9**).

5.3 Titan Integrity Campus, Bengaluru, India

Project Titan Integrity Campus, Bengaluru, India, 2017
Area 6.5 acre
Architect Mindscape Architects/ Landscape Architect: One Landscape

5.3.1 Introduction

Titan Integrity Campus, in keeping with its vision of creating elevating experiences for the people it touches and significantly impacts the world it works in, wanted to create a safe and sustainable property for its manpower within the urban agglomeration of Bengaluru. Titan Integrity field is placed on 6.5-acre site that contains a waterbody on the eastward and a road toward the north. The design has a very unique attachment with the site and the adjoining lake (as shown in **Figure 10**).

Figure 9.
Indoor outdoor experience, Christian elementary Jan van Nassau school.

Figure 10.
Titan trinity campus, Bengaluru.

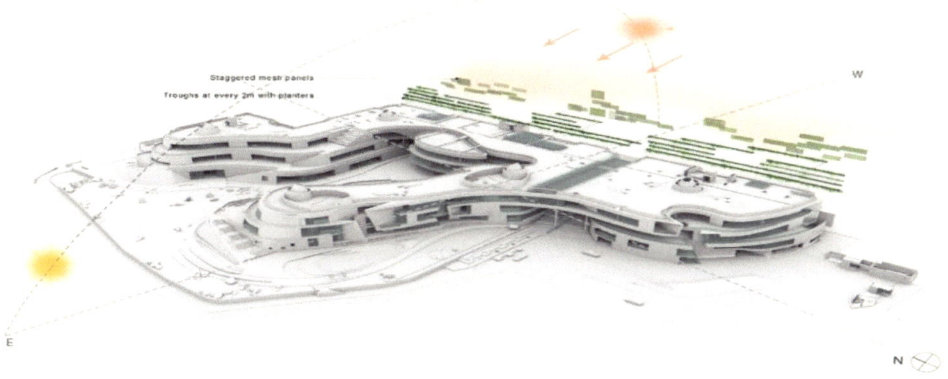

Figure 11.
Conceptual view (source: https://mindspacearchitects.com).

5.3.2 Concept

The aim was to build a campus that retains natural beauty and focuses on energy and resource conservation while enhancing the quality of life of its employees. The aim is of utilizing each potential view and feature to the lakeside makes any user a spectator of this serene setting. Lake and green spaces create an ample microclimate, thus minimizing mechanical cooling (as shown in **Figure 11**). Also, green terraces are an extension of indoor building areas that enable one to work outdoors and stimulate interaction amidst flora and fauna.

5.3.3 Spatial quality

Organization of spaces: Office building with all its ancillaries is proposed around this bio lake. Building is oriented with longer sides facing North–South to bring in glare free natural light. Porosity in planning and form allows continuous movement of breeze with wind tunnels creating venturi effect. Common areas are open and non-air-conditioned.

Porosity in structure allows continuous movement of air with wind tunnels making a venturi effect. The design is in such a way that each department has its own zonal area but is well connected to the various departments through voluminous atria, which brings in light and allows hot air to escape. The depth of area is perceived in such a way that the entire office space is filled with daylight as long as the sun allows, thus reducing the use of artificial light.

Use of landscape elements: The bio-lake alters the micro-climate and makes the place cooler than it would have been, through evaporative cooling, and also by reflection of incoming solar rays. Being quite a large body of water, it absorbs a lot of heat during the day, only to release at night. In total, 405 trees and shrubs, such as Mahagony, Champaka, and Termianila were planted on site, in 2017, and over the years, the vegetation has taken over the building, gradually making it a part of the landscape. The three-floor structure has terrace gardens at every level, which are reminiscent of rice fields, and are connected through stairs inducing a feeling of elevated ground. These terraces also provide insulation to office spaces below, thus reducing electric load, resonating with the idea of a sustainable building.

5.3.4 Indoor-outdoor experience

The design has a very special connect with the site and the adjoining lake. The idea is of exploiting every view possible to the lakeside, makes any user a spectator of this serene setting (as shown in **Figure 12**).

5.4 Googleplex, California, USA

Project Googleplex, California, USA, 2004
Area 26 acre
Architect Architect: Clive Wilkinson Architects.
Landscape Architect: Sasaki, Walker and Associates

5.4.1 Introduction

The design of Googleplex creates a strong identity for the campus and provides a much-needed civic space, blurring distinctions between the private and public realms.

5.4.2 Concept

The primary vision was to merge the idea of workplace with the experiences found in an educational environment into a new way of working and maintenance of an edge. The campus is designed to mimic the loosely structured nature of a university field. Clive Wilkinson Architects found that the nature of the company as well as the existing building complex was perfectly suited to merge the idea of workplace with the experiences found within an educational environment (as shown in **Figure 13**).

5.4.3 Spatial quality

Organization of spaces: The existing opportunities were pushed to a new level to better serve the overall goals of the community. The master plan for the entire

Figure 12.
Indoor outdoor experience—Titan integrity campus.

Figure 13.
Googleplex office view.

Figure 14.
Spatial configuration and circulation.

campus incorporates the language of a campus: outdoor sports activities, food, a common, and a park. The resultant building setup follows a straightforward distribution of work neighborhoods on a "Main Street" circulation set up (as shown in **Figure 14**). All shared resources are settled on this street and vary from meeting rooms, to technical school speak areas, to micro-kitchens and library lounges. The campus' four buildings are organized to make internal courtyards, connected to at least one another by glass bridges emphasize a way of community. A multipurpose amenities building is at the center of the field within the main court.

Use of landscape elements: At the lower level of the park, a brick plaza provides for frequent concerts and civic gatherings. The plaza's daring, stripy pattern continues up the slope through a series of terraces and shallow pools. Rows of cherry trees and fern provide color and texture of the composition. The presence of water suggests the fluid boundary between the park and field. Public art has been used extensively throughout the site (**Figures 15** and **16**).

Figure 15.
Indoor comfort spaces—Googleplex.

Figure 16.
Outdoor view—Googleplex.

5.4.4 Indoor-outdoor experience

A sinuate yellow brick path ties the campus' three gardens. The contemplative East Garden's circular mounds echo the nearby hills seen within the distance on the far side. The Central Garden acts as a congregation place for the whole community. The West Garden is dedicated to recreation, whereas the tiny low, elegant, rectangular garden contains a boccie court. Natural ventilation is key to all the spaces in order to negate the necessity for mechanical instrumentality.

6. Conclusions and discussions

6.1 Learnings and insights

The detailed deliberation above for both School Campus and Workplace provides a fresh insight to architects and policymakers for designing the spaces as well as formulating bye-laws. The importance of integration of Green Spaces within and around the built environment is witnessed through these case studies. Hence, even small efforts by policymakers for strengthening the bye-laws definitely bring cognitive development for the children at School Campus as well as reduce the urban stress and enhance the productivity at Workplace.

Moreover, architects and policymakers should also put an effort to bring the associative factors such as symbolic, experiential, active, and social spaces while designing School Campus as mentioned in the Healthy Academic Green Space framework responsible for physical, mental, and social well-being, which act as three pillars for cognitive development in children. This also helps in the formal stage development, preoperational

School Campus		
Benefits	**Case Study 1**	**Case Study 2**
	The Gyaan Centre, Jaisalmer, India	Delta-Plantsoen, Netherlands
Psychological benefits	The space within the Centre periodically acts as a marketplace to share the women's creations. Such place gives the tourists opportunities to venture to the nearby dunes to experience the sunset and harness the locational criteria of the Centre.	Bringing a natural playscape like this into the heart of the neighborhood increases children's daily contact with nature, an important factor for a healthy childhood.
Ecological benefits	The elliptical shape of the structure helps in creating a cooling panel of airflow in the hot and dry environment. The wall permits air to flow through the building and keep the Sun and sand out.	Trees were selected based on their morphological characteristics to serve specific functions. A small hollow, which holds rainwater act as a sustainable drainage system in the courtyard.
Social benefits	Renowned female artists, designers and education advocates are invited to create artwork, host events, and present installations. Students also learn to read, write, and develop ancient skills distinctive to the region.	A small pavilion stores playing equipment, which can be used by the local children. Sports equipment are also available for the local sports coaches.

Table 1.
Comparative analysis of case studies: School campus.

Workplace		
Benefits	**Case Study 3**	**Case Study 4**
	Titan Integrity Campus, Bengaluru, India	Googleplex, California, USA
Psychological benefits	Green terraces not only permit one to figure outdoors and stimulate interaction amidst the flora-fauna, but also offer trails to relish leisurely walks. The component of surprise that changes because of the change in the atmosphere around revitalizes the senses.	The garden areas of varied scales are sited throughout the campus and act as quiet pondering spots for workers. Instead of offering employees' traditional cubicles and conference rooms, the campus is set up with fluid spaces.
Ecological benefits	The terraces affect the micro-climate on the site, along with the water in the bio-lake. The landscaped terraces covered in grass, insulate the building, thereby reducing the heat load and creating comfortable conditions. Large trees provide natural shading and make the outdoors comfortable during hot summers.	The landscaped exterior offers environmental sustainability and a low-water landscaping plan. Natural light and fresh air are intelligently incorporated in the design. Perimeter bioswales and bioretention zones collect stormwater runoff for onsite treatment.
Comfort	A green wall on the western façade of the building shields the usable areas from harsh western sun. While the green buffer zone in between green wall and usable areas cuts off the radiation, the terraces conjointly offer insulation to workplace areas below.	The work areas for cryptography engineers, accommodates 3–4 workers where they can have productive work singly as well as collaboratively. The rooms are set on exterior of the building to welcome natural light.

Table 2.
Comparative analysis of case studies: Workplace.

and operational stage development as well as sensorimotor stage development in children. The learnings of the case studies as mentioned in **Table 1** are indicating the same.

The associative attributes such as energy efficiency, work-leisure congregation spaces, green-blue spaces in its various forms and scales should also be incorporated for better competence and productivity of the employees (as mentioned in **Table 2**). The integration of these spaces not only improves attention and memory, but also reduces the number of thoughts that lead to depression. Additionally, it also encourages physical activity, which can be considered as a valuable long-term investment in improving health.

Author details

Prashanti Rao*, Somaina Islary and Kapil Natawadkar
School of Planning and Architecture, Vijayawada, India

*Address all correspondence to: prashanti_swe@spav.ac.in

References

[1] Moss, Design Case Study: A Biophilic Workspace, Moss Amsterdam, 07-Oct-2021. [Online]. Available from: http://moss.amsterdam/2021/10/07/transforming-a-conventional-workspace-with-biophilic-principles/ [Accessed: February 10, 2022]

[2] Unlu A, Erincik E, Fitnat C, Mehmet S, Ervin G, Ashkan M. Interface of indoor and outdoor space in buildings a syntactic comparison of architectural schools in Istanbul. In: Proceedings of the 7th International Space Syntax Symposium Edited by Daniel Koch, Lars Marcus and Jesper Steen. Stockhol: ResearchGate; 2009. p. 132

[3] Rugel E. Green Space and Mental Health: Pathways, Impacts, and Gaps, Vancouver National Collaborating Centre for Environmental Health (NCCEH). 2015

[4] Cognitive Development—Stages, Meaning, Average, Definition, Description, Common problems, Healthofchildren.com, 2022. [Online]. Available from: http://www.healthofchildren.com/C/Cognitive-Development.html [Accessed: February 10, 2022]

[5] Bochart S. Six Principles of Biophilic Design—Biophilic Design Elements. Shepley Bulfinch. [Online]. 2014 [Accessed: February 10, 2022]

[6] Clatworthy J, Hinds J, Camic PM. Gardening as a mental health intervention: A review. Mental Health Review Journal. 2013;**18**(4):214-225. DOI: 10.1108/mhrj-02-2013-0007

[7] Kaplan D. Online teacher training of cognition and learning in education. Psychology. 2017;**08**(03):373-386. DOI: 10.4236/psych.2017.83023

[8] Cherry K. Jean Piaget's life and contributions to psychology. Verywell Mind. 2020;**8**:362 [Online]. Available from: https://www.verywellmind.com/jean-piaget-biography-1896-1980-2795549 [Accessed: February 10, 2022]

[9] Foellmer J, Kistemann T, Anthonj C. Academic greenspace and well-being—Can campus landscape be therapeutic? Evidence from a German university. Wellbeing, Space and Society. 2021;**2**:100003. DOI: 10.1016/j.wss.2020.100003

[10] Markevych I et al. Exploring pathways linking greenspace to health: Theoretical and methodological guidance. Environmental Research;**158, 2017**:301-317. DOI: 10.1016/j.envres.2017.06.028

[11] Schutte A, Torquati J, Beattie H. Impact of urban nature on executive functioning in early and middle childhood. Environment and Behavior. 2016;**49**(1):3-30. DOI: 10.1177/0013916515603095

[12] Khazan O. How nature makes kids smarter. The Atlantic. 2022;**37**:18 [Online]. Available from: https://www.theatlantic.com/health/archive/2015/06/green-spaces-make-kids-smarter/395924/. [Accessed: February 10, 2022]

[13] Whitely O. 8 Benefits of Green Office Spaces: One On Whiteley, Oneonwhiteley.co.za. [Online]. 2017 Available from: https://www.oneonwhiteley.co.za/blog/8-benefits-of-green-office-spaces/ [Accessed: February 10, 2022]

Chapter 4

Functionality and Value of Green Infrastructure in Metropolitan Sprawl: What Is the City's Future? A Case Study of the Bogotá-Sabana Northern Region

Vasco Barbosa, Mónica Marcela Suárez Pradilla,
Juan Eduardo Chica-Mejía and Jairo Eduardo Galvis Bonilla

Abstract

From the planning perspective, green infrastructure emerges as an effective strategy to address the effects of climate change and its negative impacts on urban growth. Consequently, the concept of green infrastructure represents a paradigm shift in the way in which green open space is conceived from the urban growth perspective. The concept of green infrastructure has featured prominently in urban planning since the late twentieth century. Thus, this study aims, firstly, to determine the functionality of green infrastructure in terms of population density; and secondly, to evaluate the effects of urban expansion plans on sustainable development of green infrastructure. The research applied as methods multispectral satellite imagery information to obtaining a Normalised Difference Vegetation Index (NDVI) and space syntax related with urban morphology. As main conclusion, in the municipality, there is a disconnection and disarticulation between public spaces and green areas, with the main uses, activities and facilities, as well as a road infrastructure having no connectivity. Furthermore, there is no evidence of biomass corridors that connect the space between the rivers, and the green areas identified through the NDVI analysis result more commonly to be none build areas, or low density edifications and rarely parks or natural corridors where biomass is meant to be preserved.

Keywords: green infrastructure, metropolitan urban sprawl, space syntax, climate change, urban development

1. Introduction

1.1 Green infrastructure: Concepts and planning

The concept of green infrastructure, as opposed to grey infrastructure, refers to open-space systems in cities. From the planning perspective, green infrastructure emerges as an effective strategy to address the effects of climate change and its negative impacts on urban growth [1]. Therefore, it is imperative to specify the scope of the green infrastructure concept, its objectives, principles and functions, in addition to the challenges stemming from the planning and land management fields in a metropolitan context. These challenges refer, on the one hand, to the necessary integration of green spaces planning in a physically and functionally interconnected system in the territory, and on the other hand, to an effective use of the ecosystem services that these spaces provide to the territory [2].

Consequently, the concept of green infrastructure represents a paradigm shift in the way in which green open space is conceived from the urban growth perspective. According to Benedict et al. [3], green infrastructure contributes to mitigate the ecological and social problems associated with new models of diffused urban growth, accelerated land consumption and progressive fragmentation of natural areas contribute to this. Benedict et al. also define green infrastructure as an interconnected network of green spaces that conserves the functions and values of natural ecosystems, hence extending benefits to the population [3]. Thus, the importance of the role of natural ecosystems in urban planning and development is evidenced highlighting the systemic vision in the urban growth process [4]. In brief, the contemporary shift in the green infrastructure concept emphasises the functions and benefits of green spaces provided to cities, particularly to those belonging to an interconnected and multi-scale system [5].

The concept of green infrastructure has featured prominently in urban planning since the late twentieth century [6]. According to Rouse et al. [7], as cited by Giannotti et al. [5], in the United States, the green infrastructure concept has been used since the 1990s, especially to refer to sustainable rainwater management. According to the same authors, in Europe, the term has started to be used more recently, focusing on ecological connectivity for the preservation of biodiversity and the provision of ecosystem services in both urban and rural territories. According to European Union (EU) public policy documents, green infrastructure should constitute a network of natural and semi-natural areas and other environmental elements, which are strategically planned, designed and managed for the provision of a wide range of ecosystem services [8].

1.2 Objectives and functions of green infrastructure

Mas et al. [1] refer to research advanced by the EU Working Group on Green Infrastructure according to which green infrastructure key goals are, firstly, the promotion of integrated territorial planning through identification of multifunctional areas and incorporation of habitat restoration policies related to land use; secondly, a better adaptation of society and territory to climate change scenarios; thirdly, the improvement of ecosystem services; and finally, a contribution to biodiversity conservation. Therefore, green infrastructure in urban systems has a multifunctional character with environmental, social and economic functions and value as productive areas and potentially increasing land values of surrounding areas [1].

Ramos et al. [2] provide a detailed characterisation of the functions related to green infrastructure in terms of their value as a network and other features associated with the territorial matrix. On the one hand, these authors stress functions linked to a green infrastructure network and the conservation of natural habitats that may be affected by land use dynamics, urban growth and the expansion of artificial infrastructures evidencing the importance of green infrastructure to re-establish

DOI: http://dx.doi.org/10.5772/intechopen.104453

the connectivity lost in natural habitats as a consequence of urban growth processes. Secondly, Ramos et al. point to the establishment of a system of public spaces for public recreational use and sustainable mobility for inhabitants. On the other hand, the authors propose functions associated with the territorial matrix, that is, firstly, environmental regulation functions against the impacts of phenomena associated with climate change and the so-called urban heat island; and secondly, provision functions related to the potential use of the green infrastructure system for urban agriculture. In short, green infrastructures provide diverse ecosystem services for urban populations, understood as benefits that society derives from such ecosystems [6, 9]. In a metropolitan context, these services range from the conservation of natural assets to the reduction of air pollution and the mitigation of climate change effects, as Manes et al. [10] analyse for the case of Metropolitan Rome. A highly relevant aspect of this association between green infrastructure and ecosystem services refers to the capacities of the former to spatialise the latter and to thus being able to establish the regional and local deficits of green infrastructures on the urban planning perspectives [4].

1.3 Green infrastructure in land-use planning and management

From a spatial planning perspective, the connectivity and multifunctionality of green infrastructure are two key aspects of public policies [11]. On the one hand, the connectivity concept highlights the value of green infrastructure networks for urban growth. On the other hand, the multifunctionality concept highlights the central role green infrastructure plays in territorial planning processes as strategic infrastructures for the sustainability of urbanised areas [2]. Therefore, from a land-use planning perspective, green infrastructure planning must compensate urban development while seeking to enhance connectivity with metropolitan natural border systems. Furthermore, as Mell [12] states, green infrastructure planning should rethink relations with natural systems on a metropolitan scale. In addition, green infrastructure planning policies must consider a multi-scalarity perspective to encourage an adequate articulation with urban centres [6]. This conceptual approach is a paradigm shift in spatial planning. According to Buzai et al. [4], urban models have historically emphasised economic, land rent and ecological aspects in terms of socio-spatial distribution of the population, but they have not made clear reference to land uses of natural components. Consequently, green infrastructure system planning appears even more valuable. Therefore, from a more holistic urban planning perspective, green infrastructure planning should include metropolitan green belts containing dispersed urban growth and linear parks alongside rivers that protect water resources, among other strategies [4]. A broad participation of diverse stakeholders, including public policymakers, firms, conservation entities and landowners is necessary to ensure the benefits of green infrastructures planning.

Giannotti et al. [5] study the 'Stgo+ Green Infrastructure' project in Santiago de Chile to understand how the issue of the climate emergency has been taken in consideration in green infrastructure planning and its level of priority among various types of stakeholders. Giannotti et al. conclude that stakeholders' highest priorities are the inequality in the accessibility of green areas and the potential benefits of green areas. Therefore, a key factor in green infrastructure planning and adaptation to the climate emergency should emphasise environmental justice and equity. Mas et al. [1] study the urban environmental corridor of the Manzanares River in Madrid concluding that green and blue infrastructure is relatively well implemented in climate adaptation and mitigation strategies at the European level. However, the authors highlight some lines of work that can be reinforced such as the promotion of integrated spatial planning approaches and methods to stimulate cross-sectoral regional cooperation, as well as the application of environmental-based measures in decision-making on issues related to climate adaptation and mitigation. Ramos et al. [2] analyse the provision of ecosystem services in Seville, Spain, which are based on the mapping of environmental multifunctionality in the Seville metropolitan area. The authors calculate a set of indicators jointly with the estimation of a synthetic index of green infrastructure multifunctionality. Ramos et al. conclude that both the articulation in a network of parks and natural areas and the enhancement of multifunctionality of the landscape are two essential pillars of the management of metropolitan green infrastructure. Moreover, the authors highlight the territorial matrix role and especially of agricultural areas as essential components of a multifunctional green infrastructure. For this purpose, Ramos et al. state that the Geographical Information System (GIS) is a powerful tool to facilitate the integration and spatialisation of green infrastructure designs [6].

From this perspective, the Chía, Colombia, in the case study of the Bogotá-Sabana Northern region, and more specifically in the case of the metropolitan city of Chía, which is analysed in this chapter, follows the line of argument on the role of green infrastructure in metropolitan growth, especially in metropolitan sprawl. The municipality of Chía is located in the northern Bogotá metropolitan region. Low-density and discontinuous suburban growth going beyond residences and also including educational infrastructure and offices is a pattern of the urban growth of the city [13, 14]. A major impact on the city's rural areas is unfolding, especially along the Bogotá and Frío rivers, which is leading to a fragmentation of green infrastructure and a reduction in its value and functionality for the inhabitants. **Figure 1** shows the overview landscape of the Chía case study.

Thus, this analysis aims, firstly, to determine the functionality of green infrastructure in terms of population density; and secondly, to evaluate the effects of urban expansion plans on sustainable development of green infrastructure. This study develops two research questions:

Figure 1.
Overview of landscape of Chía. Source: Authors.

How do urban green infrastructure systems contribute to urban vitality and sustainable growth in minor cities of metropolitan areas?

Does metropolitan sprawl encourage better proximity and functionality of urban green infrastructure?

Through this study, the implications and results could be used by policymakers, public officials and other stakeholders in urban planning to make informed decisions regarding strategies for the protection of green infrastructures and their adequate articulation with residential and central areas of the territory. Thus, contributing to conditions of balance between urban development and the basic environmental system of the territory.

2. Methodology

The municipality of Chía is located in the northern conurbation of the city of Bogotá (**Figure 2**), with which it has intense functional and residential relations that induce a high and continuous connectivity and exchange of products and services [15]. Chía has experienced extensive population growth due to the social, economic and functional processes generated by the urban expansion of the city of Bogotá within the boundaries of the municipality [16]. The population growth rate in Chía during the period between 1993 and 2005 was 113.2% and decreased to 35.65% between 2005 and 2018, evidencing that the expansion was lower during this period. However, due to the high cost of the square metre in Bogotá, it is likely that the population growth rate will be constant over the coming years [17].

The municipality is made up of nine hamlets located around an urban centre in which the socio-economic strata 3, 4 and 5 are mainly found [18]. According to the Colombian socio-economic stratification system, strata 1 pertains to the lowest level and strata 6 to the highest level in terms of socio-economic development. In Colombia, the urban areas of the municipalities have grown the most. Chía in 2018 showed a growth of 79.8%, four points above the national percentage average of 75.5%. The remaining 20.2% is located

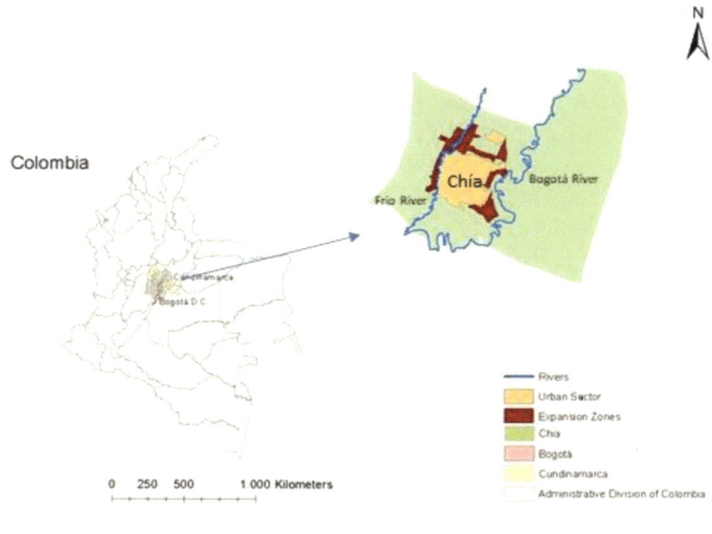

Figure 2.
Location of the municipality of Chía. Source: Authors.

especially in four of the nine hamlets of the municipality, Bojacá, la Balsa, Yerbabuena and Fonquetá, which evidences the de-ruralisation of several sectors [19].

The growth of urban land in Chía over the last 70 years is approximately 1000 ha, 55.8% of which belongs to the consolidated urban area. The urban land is distributed as follows: housing 69%, free land 18%, industrial land 5.62%, commercial land 1.27%, equipment 3.81% and crops 1.72% [20]. The municipality currently has two urban perimeters, the traditional one with an area of 590 ha, where the most important facilities are located, and a developing sector of approximately 40.5 ha [21].

This research seeks to identify the green corridors (GCs) found in the municipality of Chía, as well as to evaluate how to include them in territorial planning, especially regarding new expansion areas and in the definition of the municipal ecological structure in order to define areas of environmental quality at local and regional levels [22]. GCs are a fundamental tool in sustainable development policies as they are part of the environmental heritage of a territory and must therefore be integrated into planning and management processes [23].

The analysis of the GC of Chía considered the following zones and areas: the mountainous system of the eastern hills; the Bogotá and Frío rivers, which generate restrictions for economic and urban development activities, and which have a 30-metre zone for flood and flood management that is part of the public space of the municipality; and the mountainous system of the Majuy and Valvanera hills, which delimit the territory to the west. In addition, the environmental protection land was considered, which corresponds to the riverbeds of the tributary streams of the Frío and Bogotá rivers, with approximately 1300 ha of green areas in which effective public space sectors can be consolidated or environmentally sustainable road axes can be defined to complete the road system of the municipality that are defined in the territorial planning [24]. **Figure 3** shows that in the north–south direction, rivers are the main axes of articulation at the municipal and regional levels and have ecological and social functionality. Furthermore, from east to west, green spaces facilitate ecological connectivity along with a water system defined by the stream that lies on the perimeter of the municipality [25].

2.1 Regarding the urban expansion of the municipality of Chía

In Latin America, the definition of urban expansion areas is approached from a land use planning perspective. In Colombia in particular, such areas are defined in the Land Use Plan to ensure orderly and structured development [26] to incorporate land management instruments, organise activities and specify infrastructure and facilities needed at different times, and so Land Use Plans are the master plans according to which projects are implemented [27]. However, the review of different documents indicates that in Colombia, there is no specific policy that regulates urban sprawl. Generally speaking, urban policy promotes compact cities [28] through laws such as the Organic Law of Urban Development, which defines reserve areas for future extensions of the city, Law 61 of 1978, which in its Article 11, confers the character of public and social interest to the utility of these sectors and which is confirmed in Article 43 of Decree Law 1333 of 1986, which also establishes some limitations specifically regarding agricultural reserves, depending on the size of the municipality and its location with respect to the main urban centre [29]. In Colombia, the socio-economic stratification is a classification of residential properties that receive public services. It is carried out mainly to charge differentially by strata for residential public services, allowing subsidies to be allocated and contributions to be collected in this area. In this

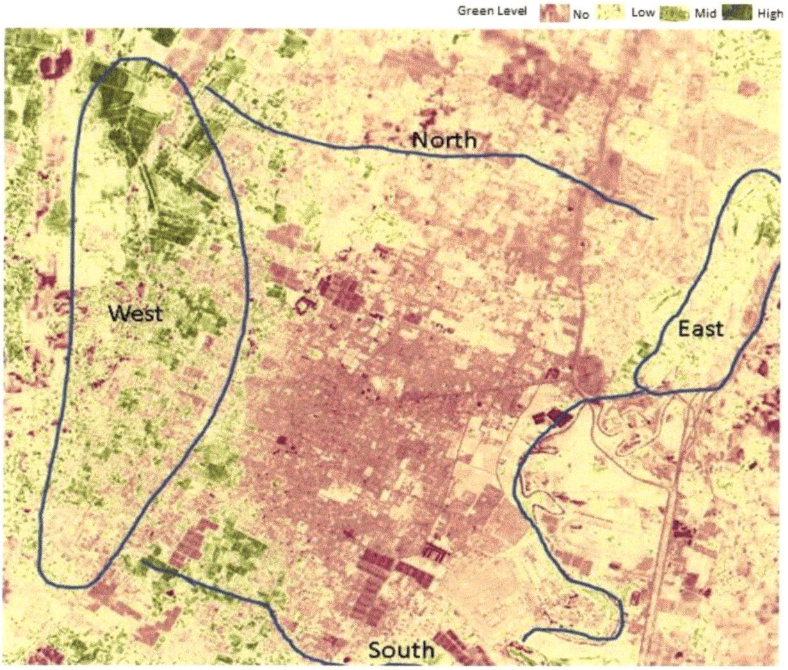

Figure 3.
Identification of green corridors. Source: Authors.

way, those with more economic capacity are made to pay more for public services and contribute so that the lower strata are helped with bill payments [30].

Law 388 of 1997 known as the Law of Territorial Development establishes that the Plan de Ordenamiento Territorial POT (Land Use Plan) must be formulated in the different Colombian municipalities. In the year 2000, the municipality of Chía proposed the first document [2] in which the current urban expansion zones of approximately 299 ha were defined. In Chía, the first planning instruments date back to the 1970s, with Agreement 16 of 1974 and 32 of 1976. In 1990, the Plan was implemented, and in 1991, Agreement 011 was developed to discourage the construction and subdivision of rural areas. In 1994, the urban planning statute for the municipality of Chía was approved with Agreement 03.

Subsequently, in 2016, in Agreement 97, a revision and adjustment of the POT were proposed considering that it had not been renewed in 12 years as indicated by legislation. Nevertheless, it was suspended in 2019, so that currently the expansion zones defined in the POT dating from the year 2000 are still in force [31].

Currently, the predominant land use is residential, so that it becomes a necessity to generate a classification of uses in the expansion zones that balances commercial and industrial aspects to expand the economic capacity of the municipality and to thus contribute to regional capacity (Kurek et al., 2020). In this process, it is also necessary to evaluate the capacity of the current road network and to enhance its responsiveness to the mobility dynamics of the municipality. Therefore, the maintenance, renovation and projection of Chía's road network are fundamental elements required to advance the diverse territorial dynamics of the municipality.

With a view to obtaining precise information related to the presence of biomass and its density in the municipality area and its relationship with urban sprawl and rural areas of Chía's territory, two main sources of information were chosen to be analysed and compared through a Geographical Information System (GIS).

The first information source is the 2018 multispectral satellite imagery information from the SAS Planet software that allows for the identification of images with red and infrared wavelength ranges, as well as visible wavelength ranges, and thus obtaining a Normalised Difference Vegetation Index (NDVI).

Using the image resources from multiple providers such as Google, Bing and ESRI, among others, the SAS Planet software facilitates downloading high-quality raster information to run multiple types of analysis over the ArcGIS platform for the purposes of identifying the greener areas of the territory of the municipality of Chía (**Figure 4**) with the Normalised Difference Vegetation Index (NDVI).

The ArcMap NDVI calculation method allows ranking the index from 0 to 100 values, identifying areas where biomass density is higher when values are closer to 100, and biomass density is low or non-existent where values are close to 0. This index rank can be obtained because '[t]he differential reflection in the red and infrared (IR) bands enables you to monitor density and intensity of green vegetation growth using the spectral reflectivity of solar radiation. Green leaves commonly show better reflection in the near-infrared wavelength range than in visible wavelength ranges. When leaves are water stressed, diseased, or dead, they become more yellow and reflect significantly less in the near-infrared range' [32].

Values from 0 to 20 are more likely to be clouds, snow, water, rocks or desert areas, as well as high-density urban areas without significantly green covered areas, where biomass tends to be critical. Values from 21 to 40 are commonly related to bushes, ground vegetation and non-dense biomass. Such values are most likely to be found in urban areas where vegetation is presented as a dispersed pattern.

Values from 41 to 60 represent very positive biomass presence. Such values are most likely to be dispersed forests, medium dense woods, where vegetation tends to be somehow greener, such as parks, forest reservoirs, mountains. Values from 61 to 80 are related to tropical and rain forest, high-density biomass where leaves are highly green and hydrated. Values equal to or higher than 81 may appear where very high density is located but are rarely obtained from NDVI analyses made from urban areas satellite images.

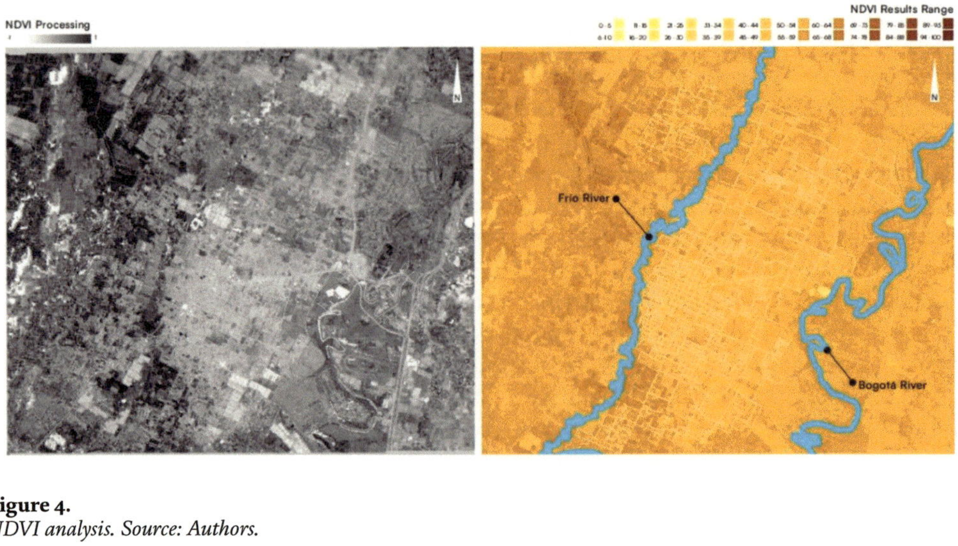

Figure 4.
NDVI analysis. Source: Authors.

The second information source selected is the 2018 version of the Colombian National Population and Housing Census, which identifies urban densities and other additional demographic information related to housing, population and life quality on an urban block scale basis.

Demographic densities were used in this case to look for possible relations between biomass density and urban areas of the municipality of Chía, where urban blocks accumulate different average NDVI values according to the presence of greener areas among them, which can be compared with the number of inhabitants per hectare residing in each urban block.

As a contribution to the study of urban green infrastructure, space syntax is applied to understand the urban morphology [33] and how the city spaces are connected to each other through the urban accessibility roads or axes. As space syntax is characterised by the study of the social and logical behaviour of space in terms of mobility, its application is considered appropriate. The aim is to deepen the knowledge of the organisation of the city in its urban axes in the context of a peripheral city and what spatial relationships may exist when compared with the green spaces of the city, that is, the urban green infrastructure.

To this end, two syntactic measures are applied given their potentiality for spatial analysis, namely connectivity and choice. Connectivity is related to the intersection of urban roads and therefore indicates areas or zones of the city where there is greater conglomeration in urban accessibility. Choice indicates the probability of a greater flow or choice of certain urban roads. To obtain these syntactic measures, a map of urban axes is drawn up to subsequently obtain the axial map of the area under study.

3. Results and discussion

The processed NDVI values show that the municipality of Chía presents biomass density common values from 35 to 70, with 50 being the most common value found, which implies that this territory has a very positive biomass presence on the analysed

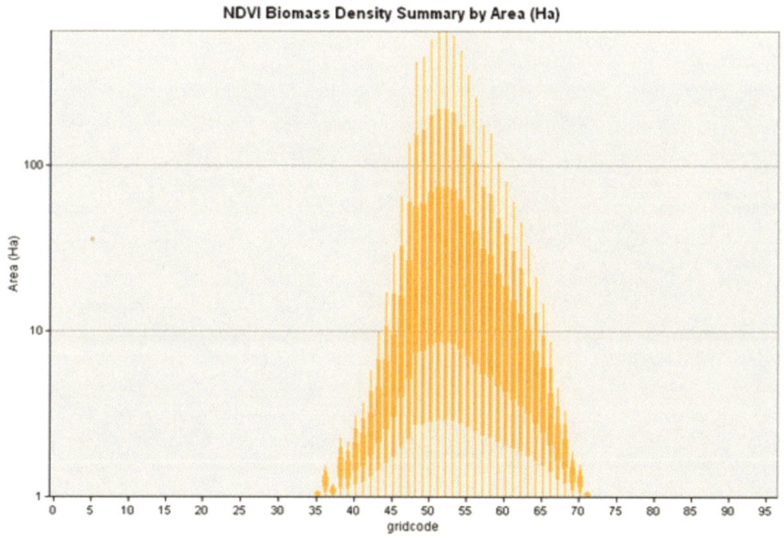

Figure 5.
NDVI analysis results. Source: Authors.

surface, with hundreds of hectares that contain different biomass elements such as trees, grass, dispersed forests and medium dense woods (**Figure** 5).

Heavily dense forests and highly dense biomass are concentrated only on the western side of the municipality, mainly on the western bank of the river Frío, where high-density biomass shows important spots that should be preserved, mainly in those zones close to main urban area.

The municipality of Chía has been growing towards the north and the south, being limited by the presence of the rivers Bogotá and Frío. This area in between the rivers displays a significant lack of biomass density compared with the western side of the municipality, where rural areas are somehow preserved with the only exception being a northwest area where the urban spot has grown at the western side of the river Frío, where it is possible to find a zone where high-density biomass can still be located among parts of the urban area.

The centralised downtown area of this municipality shows medium urban density patterns with values between 245 and 669 inhabitants per hectare, where NDVI values for the urban blocks contained in this area register values ranging from 12 to 43, indicating that there is no significant presence of biomass (**Figure 6**).

By comparing the grid code result from the NDVI analysis averaged for each urban block, and the values for urban density, it is possible to visualise that the larger portion of inhabitants reside in areas where NDVI values range from 42 to 52, mainly in the suburban areas of this municipality where the urban sprawl model has been intensively used (**Figure** 7).

The expansion areas of this municipality are located at the western, northern and eastern bounds of the main urban area and almost completely contained by the above-mentioned rivers, so that these planned expansion areas extend at least towards the northern and eastern bounds. The expansion areas lack highly dense biomass elements or significant woods or forests, and they continue the patterns that are visible in the main urban areas where there is no evidence of any strong biomass connection between the Frío and Bogotá rivers.

On the contrary, the westbound planned expansion areas indeed display a significant presence of biomass inside the marked polygons, which is a consequence of

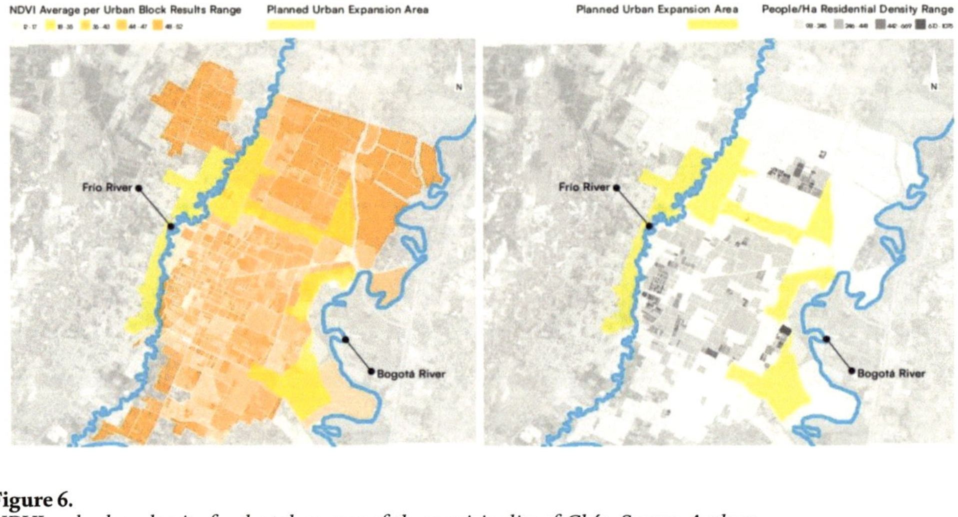

Figure 6.
NDVI and urban density for the urban area of the municipality of Chía. Source: Authors.

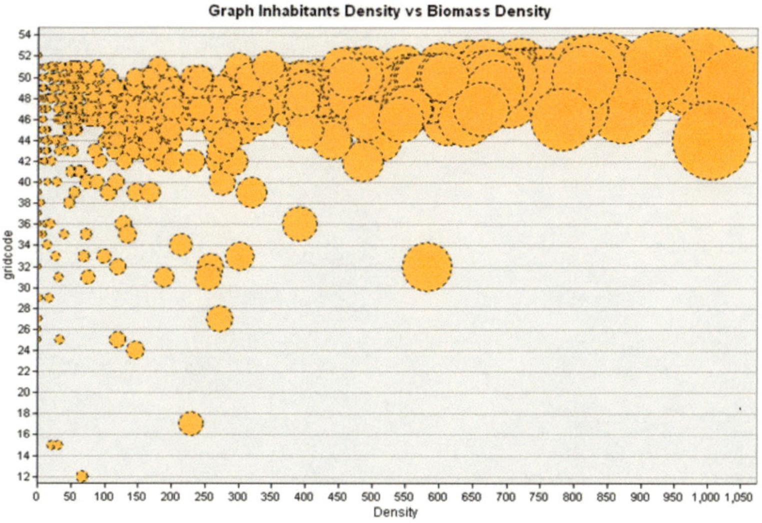

Figure 7.
Urban density vs. biomass density. Source: Authors.

the existence of some preserved rural areas where high-density biomass elements are present, which is a situation that allows us to conclude that those elements are in imminent risk of disappearance when expanded urban areas get built on the west banks of the river Frío (**Figure 8**).

By applying space syntax, we obtain the axial maps of the urban area of Chía with the connectivity and choice measures, including the respective expansion areas defined by the POT to facilitate the understanding of the results obtained (**Figure 9**). Regarding connectivity, the road with the highest values is located diagonally in the urban area and therefore away from the urban green area. Two other urban roads with high values are also located longitudinally, with the first one on the south side not coinciding with any consolidated green area, and the second one being located close to the northern expansion area. The zone near the expansion area to the east results

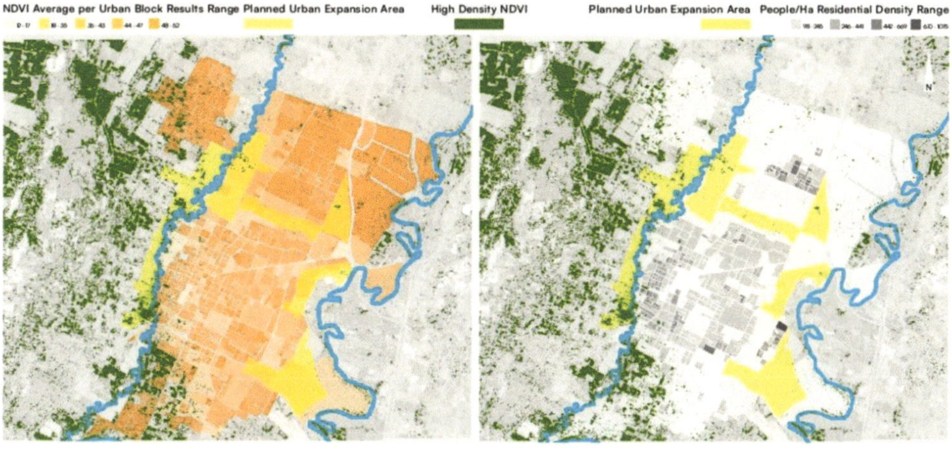

Figure 8.
High-density biomass, NDVI and urban density for the urban area of the municipality of Chía. Source: Authors.

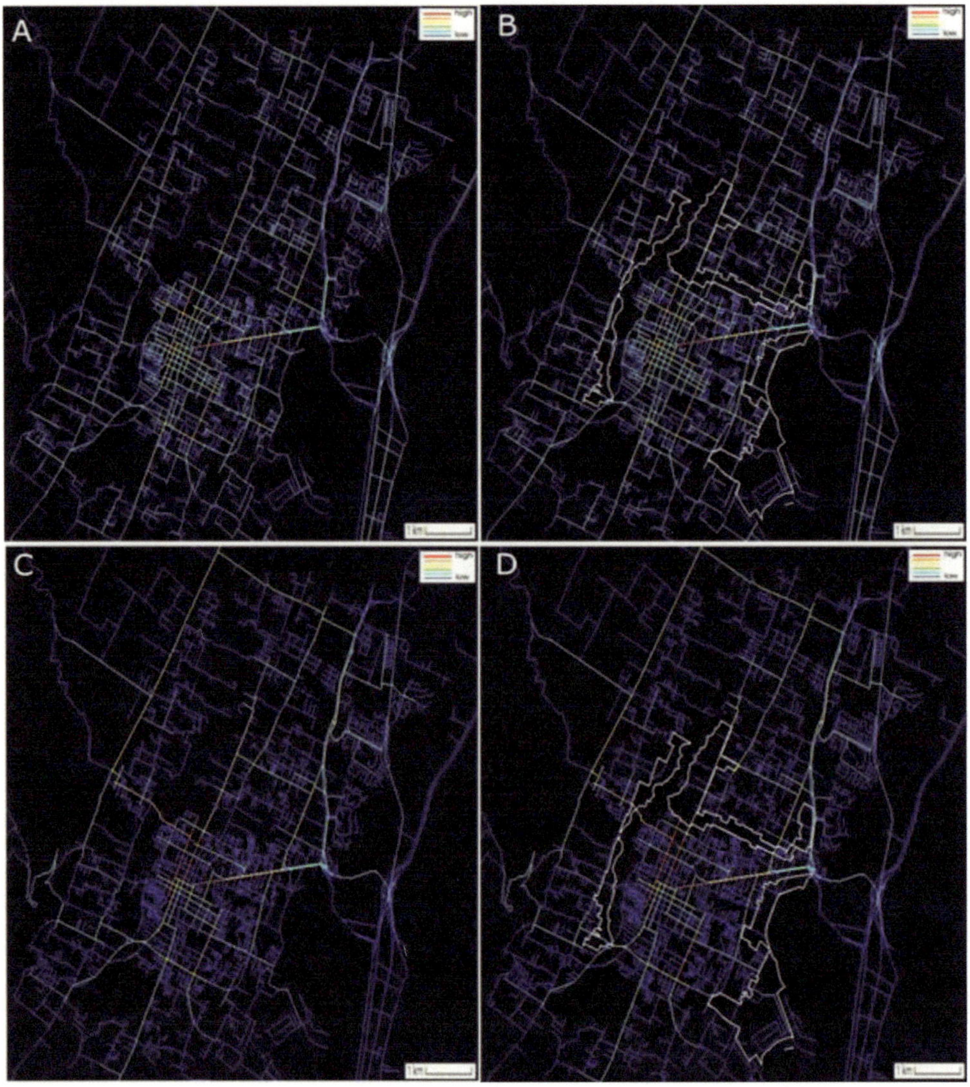

Figure 9.
Axial maps of Chía's urban space with the spatial variables: Connectivity (A), connectivity with expansion area (B), choice (C), choice with expansion area (D). Source: Authors.

from urban roads with low connectivity values. This zone with the highest concentration of green area, when analysed from the perspective of urban accessibility, results in a zone of low urban connectivity, that is, an area not inserted in the consolidated urban space, but rather in the peripheral space of the city.

On choice, the results show that two roads obtain higher values, which are defined by two longitudinal axes located near the expansion area to the North, where the green density is higher. One of these roads crosses the expansion area. However, there are no other roads that have a high probability of being elected in the urban accessibility that coincide with the consolidated green area.

In terms of environmental benefits, social well-being and economic development of communities, the importance of green infrastructure is unquestionable.

Nevertheless, the relationship between urban growth and green infrastructure location is a key factor contributing to urban vitality and sustainable urban growth. In metropolitan regions, areas remaining between the central city and other historic or emerging cities are decisive for an appropriate balance between green and built infrastructure. This can be explained, firstly, by green infrastructure contributing to a balance in urban development by mitigating negative externalities produced by urbanisation, such as pollution; secondly, by green infrastructure being fundamental for the protection of aquifers or natural areas that are highly vulnerable; and thirdly, by the fact that green infrastructure can consist of areas for urban agriculture, contributing to the population's food security.

In this study, the proximity level between green infrastructure density and population density is assumed as a key indicator of an adequate balance between urban growth and green infrastructure. This relationship may be determining the greater or lesser degree of population accessibility to these types of infrastructure.

In the case study of the Bogotá-Sabana Northern region, and more specifically in the case of the metropolitan city of Chía, the intense sprawl and emerging polarised growth do not seem to be contributing to the generation of a solid green infrastructure system as is evidenced by the relationship between low population density and low density of green infrastructure. However, the urban expansion public policies do not seem to be clearly aimed at preserving the conditions of natural green infrastructure, especially around the Frío River, which flows alongside the municipality at the western boundary of the city's historic urban centre. In general terms, the largest urban expansion areas are found along the banks of the Frío river. These expansion areas are planned with a medium-density urban form, following the compactness of the historical city. Although there are still no precise definitions of planned green and built types of infrastructure in urban expansion areas, principles of green infrastructure connectivity and functionality as a system do not seem to be guaranteed, which should be a priority of public policies for planning urban growth.

Urban growth policies in metropolitan territories must also be approached from a multi-scale perspective given the reciprocal impacts between natural and built types of infrastructure. Therefore, a solid green infrastructure should be a planning strategy that enhances the resilience of cities and provides ecosystem services to their communities.

Urban vitality and functionality promote and guarantee urban quality from the citizenship perspective. In urban development, green infrastructure takes on a relevant role and considering the results obtained by space syntax, urban morphology indicates possibilities of correction in urban design and planning. As an urban planning strategy, a concentration of green area in the expansion zone to the east near the Frío river can be envisaged. However, it seems incomprehensible that this area is not defined by nearby roads with average connectivity values to facilitate the approach and spatial integration in the city.

4. Conclusions

The growth of the municipalities of the so-called metropolitan region of Bogotá has been informal and peripheral, especially given the fact that the different characteristics that define such municipalities are not adequately and coherently articulated, which has affected the ecological structure, that is, the network of spaces and corridors in which the ecological and biodiversity processes that constitute the green space of the municipalities are developed. Accelerated urban growth has generated a great territorial imbalance associated with the challenges of environmental sustainability and climate change, which are essential elements of urban planning.

Given its location, the economic and mobility dynamics due to the increase in new homes are constant. Thus, the housing typologies that are developed are a function of the economic capacity of the new residents. Therefore, in the urban centre, the socio-economic strata 3 and 4 predominate with houses having fewer square metres and being built in height, while in the villages the population belongs to strata 5 and 6, and houses are built under a dispersed housing typology that responds to a low-density planning model.

Currently, in the municipality, there is a disconnection and disarticulation between public spaces and green areas, with the main uses, activities and facilities, as well as a road infrastructure having no connectivity and little accessibility (pedestrian, bicycle and vehicular) with the new urban developments. Considering the provisions of the POT and the fact that the urban growth of the capital city points to the north, it is likely that this dynamic will not only be maintained, but will be accelerated, since Chía has already defined expansion areas that should be developed in the future.

The urban area had growth among the two main rivers in the area, blocking any kind of ecological connectivity among them. Furthermore, there is no evidence of biomass corridors that connect the space between the rivers, and the green areas identified through the NDVI analysis result more commonly to be none build areas or low-density edifications and rarely parks or natural corridors where biomass is meant to be preserved.

The fact that the main zone with the highest density of urban green area does not include any roads with medium connectivity value in the urban morphology allows us to conclude a weak urban integration. Moreover, we can observe the decrease of urban functionality and therefore the consolidation of green infrastructure understood as a spatial element of urban development and benefit for the quality of life in the city. Urban vitality and functionality promote and guarantee urban quality from the citizenship perspective. In urban development, green infrastructure assumes a relevant role, and taking into account the results obtained by space syntax, urban morphology indicates possibilities of correction in urban design and planning.

As an urban planning strategy, a concentration of green area in the expansion zone to the east near the Frío river can be observed. However, it seems incomprehensible that this area is not defined by nearby roads with average connectivity value to facilitate the approach and spatial integration in the city.

5. Study limitations and future research lines

The most significant limitation of the study is the existence of open data and the fact that these data are not up to date; moreover, the scarcity of information in the land use plan is due to its lack of updating in relation to what is established by the municipal regulation. This may condition the development of the research process. The built structure registration data also condition the study and the understanding of this data with the urban green infrastructure, besides the scarce information that exists in the land use plan. Possible research future lines can be considered the study of urban suburbs or urban sprawl boundaries, given their built-up pressures. The connectivity analysis of the urban green infrastructure in the metropolitan region of Bogotá is also important due to the new territorial planning instrument at regional scale; therefore, its incorporation in the urban planning exercise of the municipalities surrounding the metropolis is relevant. Finally, we highlight the need to insert the urban green dimension of the municipalities in public policies, which makes up a line of research with potential in relation to climate change.

Acknowledgements

The authors are grateful to the editor of the book, as well as the reviewers for their important comments and suggestions. The authors wish to express their thanks to the Faculty of Engineering at Universidad de La Sabana, to the Department of Civil Engineering at Colombian School of Engineering Julio Garavito, to the Architecture and Habitat Department at Jorge Tadeo Lozano University and to the Research and Urban Studies Creación Urbana SAS.

Conflict of interest

"The authors declare no conflict of interest."

Author details

Vasco Barbosa[1,2,3]*, Mónica Marcela Suárez Pradilla[4], Juan Eduardo Chica-Mejía[5] and Jairo Eduardo Galvis Bonilla[6]

1 Faculty of Engineering, University La Sabana, Chía, Colombia

2 proMetheus, Polytechnic Institute of Viana do Castelo, Portugal

3 Department of Sociology, University of Johannesburg, South Africa

4 Centre for Road and Transport Studies, University Colombian School of Engineering Julio Garavito, Bogotá, Colombia

5 Architecture and Habitat Department, Jorge Tadeo Lozano University, Bogotá, Colombia

6 Research and Urban Studies Creación Urbana SAS, Bogotá, Colombia

*Address all correspondence to: vasco.barbosa@unisabana.edu.co

References

[1] Mas F, Cortés F, Molina B. Infraestructuras verdes y azules: estrategias de adaptación y mitigación ante el cambio climático. Revista Digital del Cedex. 2018;**191**:105-112

[2] Ramos J, Hurtado C. Análisis de servicios ecosistémicos para la configuración de una infraestructura verde en el área metropolitana de Sevilla. ACE: Arquitectura, Ciudad y Entorno. 2021;**16**(46):1-21. DOI: 10.5821/ace.16.46.9884

[3] Benedict MA, McMahon E. Green infrastructure: Smart conservation for the 21st century. Renewable Resources Journal. 2002;**20**(3):12-17

[4] Buzai G, Baxendale C. Modelos urbanos e infraestructura verde en ciudades de América Latina. Análisis en la ciudad de Buenos Aires. In: Santana MV, Hoyos G, Santana G, Zepeda F, Calderón J, editors. Resiliencia, territorios y gobernanza. 1st ed. Toluca: Universidad Autónoma del Estado de México; 2019. pp. 105-137

[5] Giannotti E, Vásquez A, Galdámez E, Velásquez P, Devoto C. Planificación de infraestructura verde para la emergencia climática: aprendizajes desde el proyecto "Stgo+", Santiago de Chile. Cuadernos de Geografía: Revista Colombiana de Geografía. 2021;**30**(2):359-375

[6] Aguilera F, Rodríguez V, Delgado M. Definició d'infraestructures verdes: una proposta metodològica integrada a través d'una anàlisi espacial. Documents d'Anàlisi Geogràfica. 2018;**64**(2):313-337

[7] Rouse D, Bunster-Ossa I. Green Infrastructure: A Landscape Approach. Chicago: American Planning Association. Advisory Service Report 571; 2013

[8] European Commission EC. Technical information on Green Infrastructure (GI). Commission Staff Working Document SWD (2013) 155 final. Accompanying the document communication from the Commission to the European Parliament, the Council, the European Economic and Social Committee and the Committee of the Regions Green Infrastructure (GI)-Enhancing Europe's Natural Capital COM (2013) 249 final. 2013. Bruselas. Available from: https://eur-lex.europa.eu/legal-content/EN/TXT/PDF/?uri=CELEX:52013SC0155&from=EN

[9] Fariña J, Naredo J. Libro blanco de la sostenibilidad en el planeamiento urbanístico español. Madrid: Ministerio de Vivienda; 2010

[10] Manes F, Silli V, Salvatori E, Incerti G, Galante G, Fusaro L, et al. Urban ecosystem services: Tree diversity and stability of PM10 removal in the metropolitan area of Rome. Annali di Botanica. 2014;**4**:19-26

[11] Feria JM, Ramos J. Naturaleza y ciudad. Perspectivas para la ordenación de la infraestructura verde en los planes territoriales metropolitanos en España. Boletín de la Asociación de Geógrafos Españoles. 2017;**74**:117-141. DOI: 10.21138/bage.2447

[12] Mell I. Green infrastructure planning: Policy and objectives. In: Sinnett D, Smith N, Burgess S, editors. Handbook on Green Infrastructure. UK: Edward Elgar Publishing; 2015. DOI: 10.4337/9781783474004

[13] Romero M. Territorios híbridos metropolitanos. In: Transformaciones urbanizadores en el sector central de la Sabana de Bogotá [Doctoral thesis].

Barcelona: Universidad Politécnica de Cataluña; 2019

[14] Jimenez Aguilar CM, Thoene U. Associativity in the Bogotá metropolitan region: Coordination challenges in a fragmented region. Area Development and Policy. 2021;**6**(4):451-469

[15] Universidad de la Sabana. Ministerio de Hacienda y Crédito Público, Findeter, Asocentro. Sabana Centro, Una Provincia Sostenible [Internet]. 2016. Available from: https://www.unisabana.edu.co/fileadmin/Archivos_de_usuario/Documentos/Documentos_Empresa_y_Sociedad/Vision_OTRI/Informe_Sabana_Centro._Una_Provincia_Sostenible.pdf

[16] Angarita Guzmán Tatiana. Procesos de Expansión Urbana en la Ciudad Colombiana Efectos de la Transformación del Municipio de Chía (2002-2012) [Internet]. Bogotá. 2015. Available from: https://repository.urosario.edu.co/bitstream/handle/10336/11687/1.019.034.788%20-%202015.pdf?sequence=1&isAllowed=y

[17] Alcaldía Municipal de Chía, Secretaría de Planeación. Documento de Diagnóstico. Tomo III. Dimensión sociocultural [Internet]. Chía. 2020. Available from: https://www.chia-cundinamarca.gov.co/2020/POT/3.%20DIMENSI%C3%93N%20SOCIOCULTURAL.pdf

[18] Contreras Ortíz Y. Estado de la Vivienda y del Espacio Público en el Municipio de Chía. Primera. Universidad Nacional de Colombia, Instituto de Estudios Urbanos, editors. Bogotá; 2017. Available from: https://repositorio.unal.edu.co/handle/unal/59058

[19] Hellen CG. Transformación de los espacios rurales en los municipios de Cota, Chía y Cajicá (1980-2006). Una aproximación teórico - conceptual. In: VII Seminario Nacional de investigación Urbano Regional. Medellín: Universidad Nacional de Colombia, Instituto de Estudios Urbanos; 2008. pp. 1-9. Available from: https://www.institutodeestudiosurbanos.info/descargasdocs/eventos/seminarios-de-investigacion-urbano-regional-aciur/memorias-vii-seminario-aciur-2008/mesa-6/187-transformacion-de-los-espacios-rurales-en-los-municipios-de-cota-chia-y-cajica-1980-2006/file

[20] Suárez Barrera Oscar Javier. Análisis Multitemporal de la expansión urbana de acuerdo con el aumento de la densidad poblacional del municipio de Chía para los años 1997, 2003 y 2018 [Internet]. 2020. Available from: https://repository.unimilitar.edu.co/handle/10654/36549

[21] Consejo Municipal de Chía. Acuerdo No.100 de 2016. Chía: Plan de Ordenamiento Territorial POT; 2016 p. 1-205. Available from: https://www.chia-cundinamarca.gov.co/POT2016/Acuerdo%20100%20POT%202016.pdf

[22] Lozano Castro JP, Saavedra zambrano L. Diagnóstico base para el análisis histórico ambiental en el municipio de Chía [Internet]. Bogotá. 2019. Available from: https://repository.udistrital.edu.co/handle/11349/15094

[23] Rodriguez Sarmiento José Aristides, Bueno Sarmiento José Mauricio, Santos Buitrago Oswald René. Propuesta para la generación de espacio público mediante el proyecto del Parque Lineal Río Fr [Internet]. Bogotá. 2018

[24] Cortés Vargas Julio Cesar, Rubio Zafra David Mauricio. Comportamiento espacial de la cobertura vegetal del municipio de Chía, Cundinamarca entre los años 1980-2012 y su relación con la conectividad ecológica del territorio. Manizales [Internet]. 2016. Available

from: https://ridum.umanizales.edu.co/xmlui/handle/20.500.12746/2881

[25] Soto Cifuentes Stella. Necesidad de una política de espacio público en los centros poblados del municipio de Chía [Internet]. Bogotá. 2016. Available from: https://repository.unimilitar.edu.co/handle/10654/15589

[26] Ardila Osorio GA, Rincón LM, Muñoz RJ. Implications of metropolization in the Cota-Chía-Cajicá Axis. Challenges facing the conformation of the Bogotá-Cundinamarca metropolitan region. Territorios. 2021;**45**:1-24 Available from: https://revistas.urosario.edu.co

[27] Contreras Ortíz Y. Los planes parciales como instrumentos de planeación y gestión en Bogotá: una mirada a la ejecución en la práctica [Internet]. Bogotá. 2105. Available from: http://ieu.unal.edu.co/images/documentos/Planes_Parciales_Instrumentos-Contreras_Y-2015.pdf

[28] Rodríguez vitta JF, Galarza JN, Montaño Murillo M, Salazar Tamayo MM. Manual de Innovación. Expansión Urbana Ordenada [Internet]. 2016. Available from: https://portalterritorial.dnp.gov.co/KitOT/Content/uploads/Manual%20innovacion%20Expansion%20urbana%20ordenada.pdf

[29] Gil León Santiago, Chacón Daniel Mateo. Estrategias de planificación para una ciudad en desequilibrio. Caso Chía Cundinamarca [Internet]. Bogotá. 2018. Available from: https://repository.ugc.edu.co/handle/11396/3492

[30] Paiba Ochoa LC, Cantillo Páez JS, Pulgarin Toro AF. Intervención a Través del Espacio Público y la Movilidad en el Casco Urbano del Municipio de Chía [Internet]. Bogotá. 2018. Available from: http://repository.unipiloto.edu.co/handle/20.500.12277/9225

[31] Sorensen M, Barzetti V, Keipi K, Williams J. Manejo de las áreas verdes urbanas [Internet]. Washington DC. 1998. Available from: https://publications.iadb.org/publications/spanish/document/Manejo-de-las-areas-verdes-urbanas.pdf

[32] ESRI. ArcGIS Desktop: Release 10. Redlands, CA: Environmental Systems Research Institute; 2011

[33] Barbosa V, Pradilla MMS. Identifying the social urban spatial structure of vulnerability: Towards climate change equity in Bogotá. Urban Planning. 2021;**6**(4):365-379

Chapter 5

Parameters for Designing Functional and Quality Pocket Open Spaces in High-Density Cities

Ruffina Thilakaratne

Abstract

This study discusses parameters that are important for designing quality and functional open spaces in high-density cities. Research is often limited to large parks and public squares; studies on open spaces in high-density cities are rare. Hong Kong is a high-density high-rise city where people live in compact living environments small as 12 sqm. In such contexts, open spaces play a pivotal role on human well-being. Hong Kong consists of many pocket open spaces that are intended for passive recreation. Elderly use these public amenities predominantly. Therefore, accessibility, safety and user comfort become significant considerations. Improving existing pocket open spaces is essential since there are no plans for new parks in old districts. This study analysed eight pocket open spaces, adopting for their qualitative attributes. Microclimatic field data, photographic analysis, shadow analysis simulation and user perception survey shed light on spatial design, comfort and functional aspects. This study contributed to knowledge by developing guidelines to promote quality and functionality of pocket open spaces in high-density cities.

Keywords: high-rise high-density, urban pocket open spaces, functionality, quality, urban vegetation, well-being

1. Introduction

This study adopted qualitative attributes of quality open spaces from [1–3] for analysing pocket open spaces. Research related to parks is often limited to large parks and public squares. Studies on pocket open space in high-density cities are rare. Most pocket open spaces in Hong Kong are smaller than 1000 sqm falling short compared with open space provision in other metropolitan cities. Hong Kong is a high-density high-rise city with urban population density exceeding 7126 per sqkm. Only 25% land area is developed due to challenging terrain conditions. There is a significant need for urban parks for improving emotional and physical well-being of residents who live in compact living environments as small as 12 sqm in high-rise buildings.

Zoning plans prescribe total percentage of open spaces per district based on population without further guidelines on location, size, required amenities, spatial quality and landscape design. Residual spaces in districts that were developed in 1970s were later transformed into public open spaces. Pocket open spaces considered for this study are district and local-level public amenities intended for passive recreation for the surrounding community. These are primarily being used by elderly daily in the community.

1.1 Case study district

A mixed-use residential and commercial district built under the 'new town development programme' in early 1970s provided the context for this study. This 6000 ha district accommodates 302,814 inhabitants with a density of 5300 persons per sq.km and a stock of 103,219 households. About 21.5% population are above 60 years of age [4]. To facilitate emotional and physical well-being of the increasing ageing population, easily accessible and safe outdoor amenities are vital. Although open space per capita 1 sqm in Hong Kong is far below compared with many other metropolitan cities, 87% residents live within 400 m and 94% within 800 m to public open spaces in this district [5]. In total, 131.84 ha are zoned as open spaces with sizes ranging from > 1 ha to 5 ha< to provide active and passive recreational facilities for the community [6]. Six out of eight pocket open spaces considered for this study are less than 1 ha. However, the total district-level open space provision is almost twice compared with recommended 20 ha for a population of 100,000 by the guidelines.

Average wind speed in most urban areas is approximately 1 m/s. Humidity levels from June to September range between 75 and 84% with average summer temperature reaching 34.5°C [7]. High humidity, hot weather, low wind velocity and poor air quality are key challenges deterring the enjoyment of outdoor life. Although predominant summer wind directions in this district are from the East, South and South-East quarters [8], street grids are aligned to South-East and North-West directions without facilitating adequate wind flow along the streets. Longer span of 90% of these pocket open spaces is aligned to North-East and South-West direction.

2. Influence of external parameters on open spaces

This section establishes criteria important for designing quality and functional open spaces based on previous research, benchmarking standards and guidelines. Gehl and Svarre [1–3] have contributed to the knowledge immensely through their comprehensive research on public spaces. Twelve quality criteria by Gehl Institute [3] emphasize on human perception and experience-associated aspects such as protection, comfort and enjoyment linked to public spaces. Scholars defined spatial and functional attributes of open spaces; as unbuilt spaces with a high proportion of natural elements [9, 10] cultural landscapes for socializing [11], spaces for human health, well-being and social cohesion [12] and areas for neighbourhood recreation [13]. This research extends Gehl's matrix [3] by incorporating spatial design, microclimatic and functional dimensions and responses from the surrounding built environment.

2.1 Role of open spaces on human well-being

Undeniably open spaces contribute to human well-being and quality of life besides creating desirable microclimates. World Health Organization [12] advocates open spaces for promoting healthy, liveable and sustainable cities. Green open spaces contribute to the sustainability of compact cities [14]. Dines et al. [15] opine that 'for most people, every day public spaces provide opportunities both as places of interaction and as places of retreat'. Scholars correlate nature connectedness with improved emotional and physical well-being [16, 17] and quality of life [18]. Pivotal attributes for creating quality and functional open spaces are proximity to where people live or work [19–21], presence of natural features [21, 22], cleanliness and maintenance [21], presence of amenities [22], sufficient park size [20] and people's participation in planning green spaces [23]. Summarizing [14] claims strategic planning, design and management as the key criteria for delivering functionality and benefits.

2.2 User comfort parameters

Environmental parameters such as air temperature, mean radiant temperature, wind speed and relative humidity surrounding open spaces affect thermal comfort in addition to personal parameters such as clothing insulation and metabolic rate. High-density cities often suffer from low levels of urban ventilation, yet significant level of shading from the surrounding built environment. Scholars disagree with the use of thermal comfort as feedback for design decisions due to subjective preferences by individuals and the absence of suitable thermal comfort indices for specific outdoor conditions [24–26]. Given the subjectivity of thermal comfort perception, Jansson [14] and Jendritzky et al. [27] recommend reliance on meteorological variables as a reasonable way of inferring thermal comfort conditions. Confirming this approach, Rose et al. [28] argues the importance of understanding influence from key environmental factors affecting outdoor thermal comfort, but the futility of design interventions to modify outdoor thermal comfort conditions.

2.3 Influence of greenery and hardscape on microclimatic conditions

Most urban areas consist of increased amount of impervious material that aggravates Urban Heat Island (UHI) effect through increased air temperature. Scholars advocate the use of high-albedo material for mitigating UHI effect [29–31]. Cool paving materials have shown promising results in lowering surface temperature due to their high solar reflectance and low heat storage properties [30, 32]. Although cool paving reduces air temperature, a study by Erell et al. [33] suggests their inadequacy in reducing radiant heat. Shading helps reduce mean radiant temperature compared with exposed open areas. Studies from Singapore, Taiwan and Tokyo that represent similar urban morphological and climatological conditions to Hong Kong report correlations between greenery and air temperature. Chau et al. [34] reports 1.3°C temperature difference between areas with greenery and their surroundings in Hong Kong urban parks. He further reported 1–2°C lower temperature inside the parks compared with 150 m away from the parks. Similar findings were reported by Nichol [35] based on a study conducted in Singapore. Using a numerical model, Honjo and Takakura [36] established 300 m as the optimum influenced distance from a 100 m diameter green area. Supporting above recommendations, Chen and Wong [37] reported positive contributions from greenery on microclimates within and surrounding two large urban parks in Singapore.

Kawashima [38] compared surface temperature on vegetation, buildings and exposed soil based on satellite imagery confirming lowest surface temperature on vegetation (1.4–2.7°C) compared with buildings (2–3.4°C) and soil (2.3–4.9°C). Santamouris et al. [30] reported similar trends on tree canopies 32.9°C, grass 35.6°C, under tree canopies 28.7°C and exposed concrete areas 40.7°C in Singapore CBD.

Even small green areas such as 40 × 60 m have shown 3°C reduction in temperatures in summer compared with outside areas [39]. These studies have established the role of greenery in lowering surface temperature and air temperature cooling the surrounding environment through shading and evapotranspiration.

Due to extreme high density in Hong Kong, impacts from building envelope albedo could be serious. Besides building masses, majority of horizontal surfaces are predominantly paved with very little soft landscape. Studies that compared thermal properties argued that grass and poly material contribute to very low amount of thermal energy, thereby reducing UHI effect compared with concrete and asphalt [40–42]. A study in Tokyo reported 2°C lower surface temperature at 1.2 m above grass compared with asphalt and concrete surfaces [43]. Similar conclusions were arrived at a study that compared grass and asphalt in Iran [37]. These studies support the role of urban greening on reducing air and mean radiant temperature through evapotranspiration in grass and reducing adverse impacts from low-albedo material such as asphalt.

2.4 Influence of sky view factor (SVF) on microclimatic conditions

Less sky exposure is attributed to less solar radiation entering the urban canyon, thereby reducing mean radiant temperature [44]. A study in Taipei city reported elevated temperature due to solar radiation absorption by unshaded hardscape areas within parks and surrounding areas [45]. The same study recommended reducing unshaded paved areas to less than 50% and to integrate at least 30% greenery and shading to alleviate negative effects. Comparison of three different shading scenarios; 4.9–9 m tall trees, pergolas at 4 m height around buildings with and without vegetation reported lowest mean radiant temperature around pergolas with vegetation [46]. Authors opine that pergolas and vegetation block longwave radiation. Ojaghlou and Khakzand [47] and Baghaeipoor and Nasrollahi [48] associated reduction in sky view factor (SVF) with reduction in mean radiant temperature. Analysing 18 600 × 600 m test sites in Hong Kong, Yuan and Chen [49] established a positive correlation between SVF and the Urban Heat Island effect. Adopting a software-based method [50] concluded a strong relationship between SVF and the temperature using areal means on a large sample area.

2.5 Qualitative parameters—Gehl's 12 quality criteria for public spaces

Gehl's 12 quality criteria combine human perception-based aspects such as protection, comfort and enjoyment articulated through pragmatic criteria (**Table 1**). Pragmatic aspects focus on accessibility, navigation within, seating options, safety concerns, impact from environmental conditions, aesthetics and how people interact. These criteria are derived from research on public spaces in European cities. Therefore, some of these criteria require modifications to suite high-density Hong Kong context.

2.6 Hong Kong Planning Standards and Guidelines (HKPSG)

Hong Kong Planning Standards and Guidelines [6] advocate sustainable and liveable open spaces through four guiding principles: 'quantity', 'quality', 'good practice' and 'vision' (**Table 2**). HKSAR Planning Department [6] recommends that open spaces should be safe, functional, accessible and usable spaces for the community,

Protection	Protection against traffic and accidents
	Protection against harm by others
	Protection against unpleasant sensory experience
Comfort	Options for mobility
	Options to stand and linger
	Option for sitting
	Options for seeing
	Options for talking/ hearing
	Options for play, exercise and activities
Enjoyment	Human scale
	Opportunities to enjoy positive aspects of climate
	Experience of aesthetic qualities and positive sensory experiences

Table 1.
Gehl's 12 quality criteria (2018) for public urban spaces (source: adopted from Gehl and Svarre [3]).

Quantity	1 sqm per person for district level and local level open spaces
Quality	high quality facilities, environmental standards and layout design in-line with users' aspirations
Good practice	provide easy access, improve facilities as a community
Vision	assessment of existing facilities in relation to location, demographics, land uses and functions and socio-economic characteristics and prospects for future developments of the district

Table 2.
HKPSG on delivering sustainable and liveable open spaces.

not just residual spaces for merely fulfilling regulations. Local open spaces are non-statutory land uses and should be at least 500 sqm in extent in urban areas intended for passive uses. Government Greening Policy aspires enhancing urban ecology by active planting and preserving and maintaining trees.

Open spaces should cater to elderly, persons with disability, children and adults in an integrated manner; therefore, safety is a major consideration in open space design with regard to location, identifiable entrances, surface texture and facilities provided. Provision of slip resistant surface finishes, adequate furniture and positioning them under shading and away from the pedestrian paths are essential.

However, these guidelines do not provide measurable criteria for designing or assessing open spaces in relation to quality, good practices and vision objectives in **Table 2**.

3. Developing spatial design and user comfort parameters for functional pocket open space design

3.1 Scale, accessibility and safety

Since the case study district was developed prior to the announcement of HKPSG [6], most open spaces considered for this study do not adhere to the guidelines. In total, 500 sqm minimum open space extent and 1 sqm per capita requirements have not been met in most instances. Majority of pocket open spaces appears to be residual spaces located amidst of commercial areas, residential blocks and adjacent to traffic roads.

Open spaces in the case study district varied from 200 sqm to 10,600 sqm, with only four out of eight open spaces being over 500 sqm. All eight urban public open spaces are located within 100 m radius from residential and commercial neighbourhoods and are well integrated into the urban fabric ensuring safety and easy accessibility (**Figure 1**). 'LEED Cities and Communities Plan and Design Standards' [51] specify 11.25 sqm per capita green open spaces within 400 m radius from residential areas.

Surrounding context consists of six to seven storeys tall old Chinese shop houses and 10–15 storeys tall public housing estates providing sense of human scale and sense of protection. Given the long operating hours of the commercial areas until 10 pm, these parks are well lit during day and night. Survey respondents appreciated the proximity, sense of safety, emotional comfort and physical well-being they receive by these pocket open spaces. Mantler and Logan [16] and McEwan et al. [18] reported positive impacts on adults spending time outdoors. Two open spaces are located adjacent to major traffic roads curtailing the enjoyment due to likely traffic accidents, exposure to noise and air pollution. These parks are completely pedestrianized; therefore, no bicycle paths are provided.

Analysing large number of open spaces in Hong Kong, Civic Exchange Hong Kong [52] opines, 'access and linkages', 'comfort and image', 'uses and activities' and 'sociability' are vital in creating quality open spaces. Overall, these parks aligned with over 60% of the 'protection' and 'comfort' criteria according to Gehl's matrix.

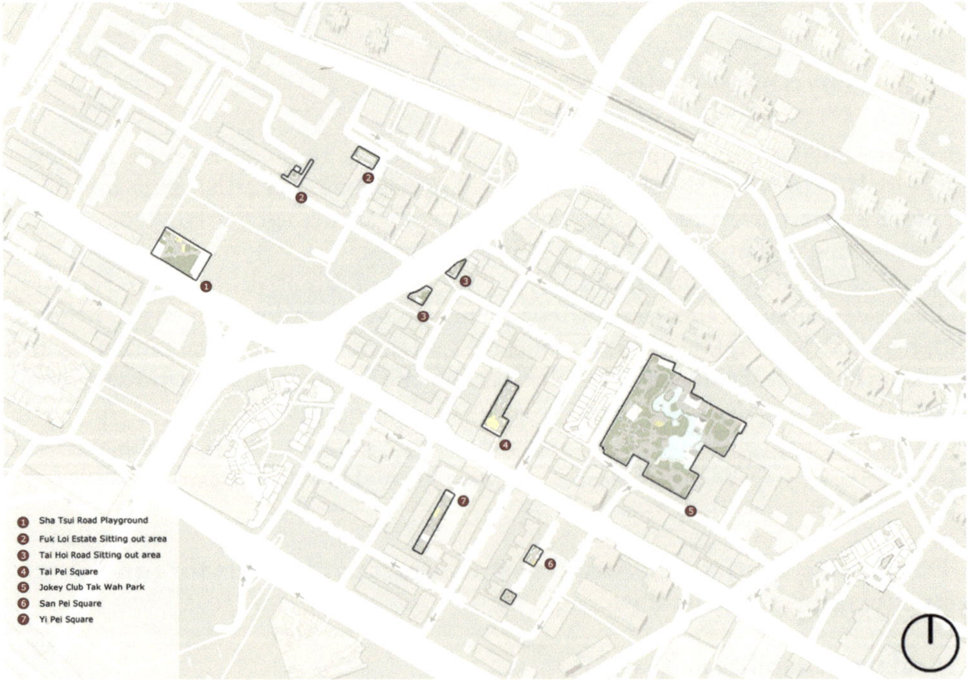

Figure 1.
Park network indicating location and orientation in relation to summer wind directions; East, South and South-East quarters.

3.2 Spatial design and quality

User survey and field studies revealed that these open spaces are primarily being used for passive recreation by senior citizens during late mornings and late afternoons. All these pocket open spaces are accessible to elderly and disabled persons. One consists of basic fitness equipment for elderly, and two consist of play elements for children. Six out of eight open spaces are predominantly made of hardscape. HKSAR Planning Department [6] stipulates minimum 70% soft landscape inclusive of 60% trees for passive recreational areas. None of these open spaces have achieved these standards as these parks were established before the guidelines were enacted. Although they are conveniently located, there are no identifiable entrances or boundaries to majority of these open spaces. Due to the compact size, navigation within was straightforward, however, lacked clear demarcations between sitting-out areas and pedestrian paths. Only the largest park is designed with defined landscape and hardscape areas, walkways and variety of sitting-out areas for small and large groups. Although seating has been the primary attraction in these open spaces, current seating arrangement is ad hoc; not all seating places are provided in shaded areas. Photographic survey confirmed users' preference for shaded areas.

Although seating has been the primary attraction in these pocket open spaces, shadow analysis confirmed random positioning of seating without aligning with shading that occurs from the surrounding built environment. Five open spaces are paved with light colour cement blocks; two with orange colour eco blocks and two open spaces surfaced with dark colour rubber mats in the activity areas. None of these open spaces have paid attention to views or focal points.

These open spaces demonstrated about 50% alignment with the 'comfort/ spatial design and quality' criteria in the matrix.

3.3 Microclimatic aspects and user comfort

Air temperature, solar radiation, relative humidity, wind speed and direction were considered as the meteorological measurements affecting thermal comfort.

Urban heat island effect caused by excessive hardscapes in the parks and surrounding building density and low urban wind speed challenge user comfort during summer. **Figure 2** presents an analysis of hardscape to greenery ratio and orientation in relation to summer wind directions; East, South and South-East quarters.

Figure 3 presents a strong inverse correlation between greenery percentage and air temperature with $r = -0.77$ supporting that increased vegetation helps reduce air temperature in these pocket open spaces. Although the temperature range was narrow, lowest temperature values are associated with greenery above 49%. Effectiveness of trees in lowering air temperature in parks and the vicinity have been reported by previous scholars [29–32].

Santillán-Soto et al. [40] reported elevated temperature levels in unshaded hardscapes in parks in Taipei due to solar radiation absorption. This study recommended that parks should be designed with less than 50% of paved areas and at least 30% of vegetation and shading. On the contrary to scholars' definition of open spaces as unbuilt spatial and functional features with high proportion of natural elements [9] and areas with permeable soft surfaces [53], seven out of eight pocket open spaces predominantly consisted of hardscape despite prevalent 2–3.5°C UHI effect in Hong Kong [54]. BEAM Plus Hong Kong green building guidelines recommend 50% or more passive open spaces and pedestrian zones to achieve thermal comfort [55]. Although these standards refer to new development projects, they provide a quantifiable criterion applicable to Hong Kong.

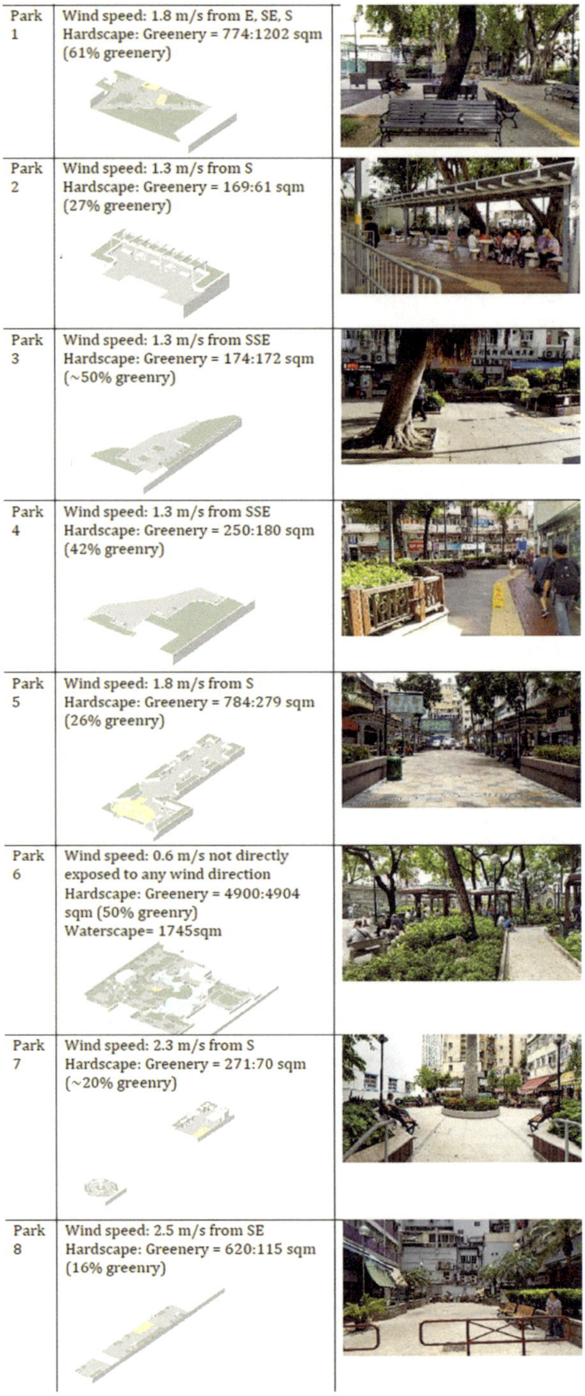

Figure 2.
Microclimatic and spatial analysis of the pocket parks.

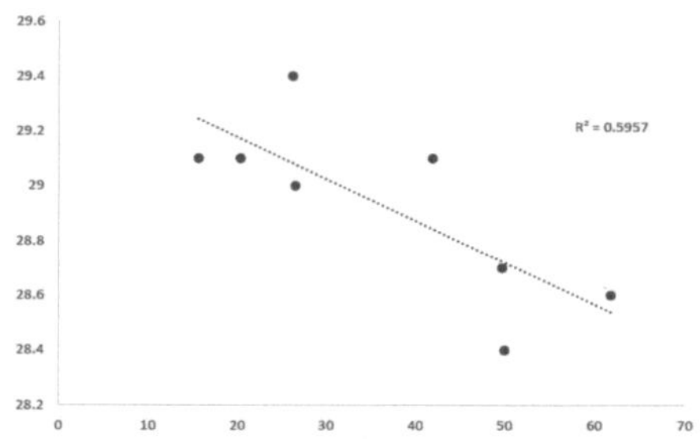

Figure 3.
Correlation between air temperature and greenery % in pocket open spaces.

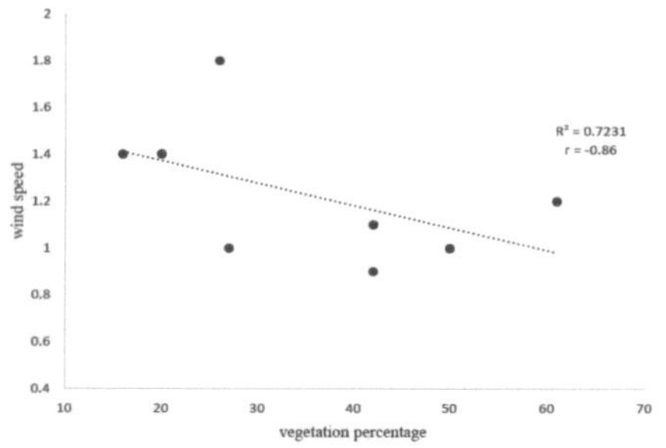

Figure 4.
Greenery % and wind speed in pocket open spaces.

Figure 4 indicates a strong inverse correlation with r= −0.86; parks with higher greenery percentages reported lower wind speeds perhaps due to friction created by hedges and shrubs. Refs. [56–59] also confirmed reduced wind speed within urban canopy due to trees. Parks with open passages diagonal or perpendicular to the prevailing wind directions reported high wind speeds.

Hong Kong experiences a warm humid summer with high relative humidity and low wind speed affecting outdoor thermal comfort. In Hong Kong at 28°C temperature, a wind speed of 0.9–1.3 m/s is required to provide neutral thermal comfort for a person in light clothing [60]. All open spaces reported above 28°C, yet with wind speeds ranging between 0.9 and 1.8 m/s presumably providing users with comfortable outdoor thermal environments.

These open spaces are surrounded by mid-rise to high-rise buildings, casting partial shadows at different times during the day (**Figure 5**). Number of scholars have associated low Sky View Factor (SVF) with low air temperature due to reduced solar radiation [39, 42–45]. This is another indication of desirable thermal comfort in these parks if seating positions are aligned with these shadow patterns. These open spaces demonstrated about 60% alignment with 'enjoyment/ microclimatic and thermal comfort aspects' in Gehl's quality criteria matrix [3].

3.4 User perceptions and expectations

Findings and guidelines from [3, 5, 6, 47] were adopted when developing user survey. Purpose of the survey was to understand users' perceptions associated with these pocket open spaces and to understand their expectations. Most frequent users are male over 50 who use these pocket open spaces 5–7 days a week: whilst users

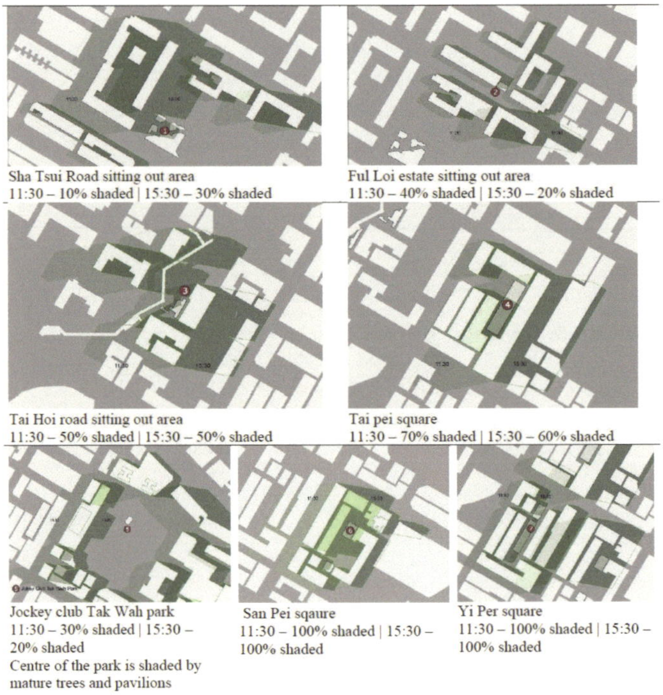

Figure 5.
Shadow analysis that compares 11:30 and 15:30 h on a sunny day.

between 30 and 40 years old use them less than 2 days a week between 3 and 6 pm. All age groups use these parks for relaxation and socializing with friends.

Users appreciate the proximity of these open spaces to their neighbourhoods and the use of slip resistant paving material providing them sense of safety, emotional comfort and physical well-being. World Health Organization [12] emphasizes the need for green spaces and their proximity; refs. [16, 18] confirm positive impacts on adults spending time outdoors.

Users' concerns include small size, over crowdedness, noise, lack of separation from traffic roads, with further emphasis on the need for improving functionality, quality and comfort-related aspects such as pavilions for resting, seating under shade, sanitary facilities, hygiene and more landscape. Further they wish for large parks, large canopy trees, colourful landscapes, natural elements and views.

4. Recommended criteria for creating quality and functional pocket open spaces

Considering the increasing ageing population in Hong Kong, public amenities should adopt an all-encompassing approach to ensure user comfort, well-being and safety. Key parameters are categorized into the following: planning, spatial and design, landscape, facilities, maintenance. Current planning regulations should focus on specifying strategic locations for positioning open spaces besides providing guidelines. **Table 3** presents recommendations.

	Aspect	Current provision	Recommendation and [relevant references]
Planning aspects	Location	• Randomly located	• Provide open spaces away from major traffic roads, yet in close proximity to neighbourhoods [3] • Define specific locations on district Outline Zoning Plans. • Demarcate clear boundaries and recognizable entrances
	Accessibility	• No radius specified other than easy access [6]	• Within 400 m radius [51]
Spatial and design aspects	Provision	• 1 sqm per capita	• Minimum 4 sqm per capita [52] • Combine scattered and small open spaces to form large parks at least 5000 sqm [6]
	Configuration	• Ill-defined	• Regular and defined space [61] • Well-proportioned [61]
	Universal accessibility	• All parks are elderly friendly	• Avoid steps [62] • Adopt ramps not less than 1:12 gradient with guardrails, and tactile paving for elderly and disables [62]
	Surface material	• Hardscape	• Soft anti-skid material and tactile paving [62] • Introduce colours to increase contrast for elderly [62] • Use permeable material or grass cover to prevent storm water stagnation • Use high-albedo material to reduce Urban Heat Island effect • Reduce concrete paving [3, 63] • Adopt combination of vegetation ground cover and cool paving or high-albedo material to reduce radiant temperature [30, 32, 33, 38, 39, 42]
	Landscape	• Not according to any guidelines	• Increase vegetation ground cover to reduce air temperature [38, 39, 42] • 70% landscape with 60% trees [6] • Large tall trees for shading and to prevent obstruction to wind paths [56–59, 64] • Colourful flowering plants [based on user survey] • Ecological continuity with native plants for low maintenance [61]
	Seating arrangement	• Randomly located and on exposed areas	• Variety of seating for small and large groups under shading [3] • Seating provided under shading • Seating away from pedestrian paths [based on user survey and shadow analysis]

	Aspect	Current provision	Recommendation and [relevant references]
Comfort	Lighting	• Well-lit during daytime • Street / park lighting and lighting from surrounding environment at night	• Sensor driven lighting to provide protection and for energy efficiency • Downward lighting to prevent nightlight pollution
	Thermal comfort	• Low wind speeds • Unshaded areas • Heat absorbing hard surfaces	• Pavilions and trees for shading [34–37] • Provide wind deflectors to divert S, SE, E wind to these parks • Refer to hardscape and landscape recommendations above
	Noise and pollution mitigation	• Some parks are bordering traffic roads	• Avoid parks adjacent to traffic roads • For existing parks install noise barriers adjacent to traffic roads • Plant trees to reduce pollutants [64–67]
	Well-being	• No attention to well-being aspects	• Create views (sculptures, arts, mural, fountains etc. to increase aesthetic appeal) • Integrate natural elements such as stones, wood, water, trees • Create variety of different zones within to cater to personal preferences • based on user survey and [68]
	Facilities	• Not provided on a consistent basis	• Drinking fountains [68], toilets • Fitness equipment • Play equipment [based on user survey]
Operational & feedback	Maintenance Feedback	• Managed by Leisure and Cultural Services Department (LCSD)	• Involve community in maintaining these open spaces • Require regular maintenance and guidelines for users • Conduct community engagement exercises when planning and upgrading open spaces [69, 70]

Table 3.
Recommended planning, spatial, comfort and operational criteria for creating functional and quality pocket open spaces.

5. Limitations and further research directions

This study revealed number of correlations that are worthy of further investigations. Impacts from shading by surrounding buildings on reducing air temperature are pertinent and beneficial to high-density cities. Although trees contribute to reducing air temperature, further studies are important to establishing tree densities or layouts that would not compromise wind speed within the urban canopy.

6. Conclusions

This study focused on the functionality and quality of public pocket open spaces in extremely high-density Hong Kong where people live in compact high-rise towers.

Eight open spaces from a mixed-use residential district that was developed in early 1970s provided the context for this study. Over 21.5% of population represents elderly above 60 years. All these open spaces have followed inclusive design principles making them elderly- and disable-friendly.

This study adopted Planning Standards and Guidelines Draft Outline Zoning Plan 2017 four principles: 'quantity', 'quality', 'vision' and 'good practice', Gehl's 'twelve quality criteria for experiencing public spaces' as the basis for qualitative analysis validated by spatial, microclimatic, shadow analysis, photographic data, user behaviour patterns and user perception surveys.

Certain aspects in Gehl quality criteria [3] were not applicable to Hong Kong context. This study contributed to knowledge by developing design guidelines to promote quality and functionality of pocket open spaces in high-density cities. In terms of open space per capita and the extent, these pocket open spaces fell short compared with other similar cities in Asia [47]. Proximity of these open spaces to residential estates and commercial developments contributed to sense of protection, accessibility and frequent usage by the elderly in the late afternoons when parks cool off.

Given the small scale, navigation within was convenient. These parks had no clear entrances and segregation between pedestrian paths and seating areas. Seating areas are randomly positioned without benefiting from shading that occurs from the surrounding tall buildings.

Majority of these parks predominantly consisted of hardscape despite prevalent urban heat island effect. However, field data supported desirable thermal comfort conditions in these parks perhaps attributed to combination of other factors such as shading from surrounding buildings, alignment to wind paths and vegetation in some parks.

Landscape, shading, seating arrangements and provision of sanitary facilities require improvement to promote user comfort and well-being. User survey confirmed findings from qualitative, spatial and microclimatic analysis.

Although these parks made no reference to any standards at the time they were developed, majority of the parks indicated considerable alignment with Gehl's quality criteria.

Acknowledgements

The author is grateful to Chu Hai College of Higher Education for research funding. Author would like to thank student assistants, Lam Ho and Yiksun Lai, for their contribution in data collection, graphics and simulation studies, Dyrus Hau, Joyce Lee, Joanna Chan and Villy Choi for conducting user surveys and Mr. R. Tan for sharing planning guidelines knowledge.

Conflict of interest

The authors declare no conflict of interest.

Author details

Ruffina Thilakaratne
Chu Hai College of Higher Education, Tuen Mun, Hong Kong

*Address all correspondence to: ruffinaat@chuhai.edu.hk

References

[1] Jan G. Life between buildings: using public space. 6th ed. Washington DC: Island Press; 2011

[2] Gehl Institute. Gehl's Twelve Quality Criteria for public urban spaces. Gehl Institute; 2018

[3] Gehl J, Svarre B. How to study public life. In: How to Study Public Life. 2nd ed. Washington DC: Island press; 2013

[4] HKSAR Census and statistics Department. Population by Census district profiles. 2016. Available from: https://www.bycensus2016.gov.hk/en/bc-dp-new-town.html

[5] Chow J. Public open space accessibility in Hong Kong, a geospatial analysis Reconnecting open space programme. Hong Kong: Civic Exchange; 2018

[6] HKSAR. Planning Department. Chapter 4—Recreation, Open Space and Greening [Internet]. 2015. Available from: https://www.pland.gov.hk/pland_en/tech_doc/hkpsg/full/pdf/ch4.pdf

[7] Observatory HKSAR. Daily Weather Summary and Radiation Level. 2018. Available from: http://www.hko.gov.hk/cgi-bin/hko/yes.pl?year=2018&month=06&day=26&language=english&B1=Confirm

[8] Ng E. Cat. A1—Term consultancy for expert evaluation and advisory services on air ventilation assessment Contract No.: PLNQ 35/2009. Hong Kong; 2009

[9] Maruani T, Amit-Cohen I. Open space planning models: A review of approaches and methods. Landscape and Urban Planning. 2007;**81**:1-13

[10] Lo AYH, Jim CY. Differential community effects on perception and use of urban greenspaces. Cities. 2010;**27**(6):430-442

[11] Law L. Defying disappearance: Cosmopolitan public spaces in Hong Kong. Urban Studies. 2002;**39**(9):1625-1645

[12] World Health Organization. Urban Green Space Interventions and Health: A review of impacts and Effectiveness [Internet]. Copenhagen: WHO Regional Office for Europe; 2017. 2022. Available from: https://www.euro.who.int/__data/assets/pdf_file/0010/337690/FULL-REPORT-for-LLP.pdf

[13] Thompson W. Urban open space in the 21st century. Landscape and Urban Planning. 2002;**60**:59-72

[14] Jansson M. Green space in compact cities: the benefits and values of urban ecosystem services in planning. Nordic Journal of Architectural Research. 2014;**2**:139-160

[15] Dines N, Cattell V, Gesler W, Curtis S. Public Spaces, Social Relations and Well-Being in East London. Bristol: The Policy Press; 2006

[16] Mantler A, Logan A. Natural environments and mental health. Advances. Integrative Medicine. 2015;**2**:5-12

[17] Herzog TR, Kaplan S, Kaplan R. The prediction of preference for unfamiliar urban places. Population and Environment. 1982;**5**(1):43-59

[18] McEwan K, Richardson M, Sheffield D, Ferguson FJ, Brindley P. A smartphone app for improving mental health through connecting with urban nature. International Journal of Environmental Research and Public Health. 2019;**16**(18):1-15

[19] Nielsen T, Hansen K. Nearby nature and green areas encourage outdoor activities and decrease mental stress. CAB Reviews. Perspectives in Agriculture, Veterinary Sciences, Nutrition and Natural Resources. 2006;**1**:1-10

[20] Giles-Corti B, Broomhall M, Knuiman M, Collins C, Douglas NK, et al. Increasing walking: how important is distance to, attractiveness, and size of public open space? American Journal of Preventive Medicine. 2005;**28**(2):169-176

[21] Gobster PH. Managing urban parks for a racially and ethnically diverse clientele. Leisure Sciences. 2002;**24**:143-159

[22] Tinsley HE, Tinsley DJ, Croskeys CE. Park usage, social milieu, and psychosocial benefits of park use reported by older urban park users from four ethnic groups. Leisure Sciences. 2002;**24**(2):1990218

[23] Huang S. The impact of public participation on the effectiveness of, and users' attachment to, urban neighbourhood parks. Landscape Research. 2010;**35**(5):551-561

[24] Givoni B, Noguchi M, Saaroni H, Pochter O, Yaacov Y, Feller N, et al. Outdoor comfort research issues. Energy and Buildings. 2003;**35**(1):77-86

[25] Chen L, Ng E. Outdoor thermal comfort and outdoor activities: A review of research in the past decade. Cities. 2012;**29**(2):118-125

[26] Brown R, Vanos J, Kenny N, Lenzholzer S. Designing urban parks that ameliorate the effects of climate change. Landscape and Planning. 2015;**138**:118-131

[27] Jendritzky G, de Dear R, Havenith G. UTCI—Why another thermal index? International Journal of Biometeorology. 2012;**56**(3):421-428

[28] Rose D, Wu H, Beyers M. Spatial and temporal computation of thermal comfort inputs in outdoor spaces. The Fifth International Symposium on Computational Wind Engineering (CWE2010). Chapel Hill, North Carolina, USA. 23-27 May 2010

[29] Santamouris M, Synnefa A, Kolokotsa D, Dimitriou V, Apostolakis K. Passive cooling of the built environment—Use of innovative reflective materials to fight heat islands and decrease cooling needs. International Journal of Low-Carbon Technologies. 2008;**3**(2):71-82

[30] Santamouris M, Gaitani N, Spanou A, Saliari M, Giannopoulou K, Saliari M. Using cool paving materials to improve microclimate of urban areas—Design realization and results of the flisvos project. Building and Environment. 2012;**53**:128-136

[31] Akbari H, Pomerantz M, Taha H. Cool surfaces and shade trees to reduce energy use and improve air quality in urban areas. Solar Energy. 2011;**70**(3):295-310

[32] U.S. Environmental protection agency. Cool pavements. In: Reducing urban heat islands: Compendium of strategies. Draft. 2012. Available from: https://www.epa.gov/heat-islands/heat-island-compendium

[33] Erell E, Pearlmutter D, Boneh D, Kutiel P. Effect of high-albedo materials on pedestrian heat stress in urban street canyons. Urban Climate. 2014;**10**:367-386

[34] Chau BC, Chau C, Chan H, Yung H. Investigating the Effects of Greenery on Temperature and Thermal Comfort in Urban Parks Research Project Title:

Investigating the effect of greenery on temperature and thermal comfort in urban parks. Hong Kong: Hong Kong Polytechnic University; 2016

[35] Nichol J. High-resolution surface temperature related to urban morphology in a tropical city: a satellite-based study. Journal of Applied Meteorology. 1996;**35**:135-146

[36] Honjo T, Takakura T. Simulation of thermal effects of urban green areas on their surrounding areas. Energy and Buildings. 1990;**15**(3-4):443-446

[37] Chen Y, Wong NH. Thermal benefits of city parks. Energy and Buildings. 2006;**38**(2):105-120

[38] Kawashima S. Effect of vegetation on surface temperature in urban and suburban areas in winter. Energy and Buildings. 1991;**15**(3):465-469

[39] Saito I, Ishihara O, Katayama T. Study of the effect of green areas on the thermal environment in an urban area. Energy and Buildings. 1990;**15**(3-4):493-498

[40] Santillán-Soto N, García-Cueto R, Haro-Rincón Z, Ojeda-Benítez S, Quintero-Núñez M, Velázquez-Limón N. Radiation balance of urban materials and their thermal impact in semi-desert region: Mexicali, México study case. Atmosphere. 2015;**6**:1578-1589

[41] Shahmohamadi P, Che-Ani A, Ramly A, Maulud K, Mohd-Nor M. Reducing urban heat island effects: A systematic review to achieve energy consumption balance. International Journal of Physical Sciences. 2010;5(6):626-636

[42] Madjidi F, Bidokhti A, Ghader S, Mansori N. Cooling and warming effects of a grass covered area and adjacent asphalt area in a hot day. International Journal of Occupational Hygiene. 2013;**5**(2):56-63

[43] Ca TV, Asaeda T, Abu EM. Reductions in air conditioning energy caused by a nearby park. Energy and Buildings. 1998;**29**(1):83-92

[44] Givoni B. Climate Considerations in Building and Urban Design. New York, Chichester, Weinheim, Brisbane, Singapore, Toronto: John Wiley & Sons Inc.; 1998. p. 480

[45] Chang CR, Li MH. Effects of urban parks on the local urban thermal environment. Urban Forestry and Urban Greening. 2014;**13**(4)

[46] Ridha S, Ginestet S, Lorente S. Effect of the shadings pattern and greenery strategies on the outdoor thermal comfort. International Journal of Engineering and Technology. 2018;**10**(2):108-114

[47] Ojaghlou M, Khakzand M. Cooling effect of shaded open spaces on long-term outdoor comfort by evaluation of UTCI index in two Universities of Tehran. Space Ontology International Journal. 2017;**6**(2):9-26

[48] Baghaeipoor G, Nasrollahi N. The effect of sky view factor on air temperature in high-rise urban residential environments. Journal of Daylighting. 2019;**6**(2):42-51

[49] Yuan C, Chen L. Mitigating urban heat island effects in high-density cities based on sky view factor and urban morphological understanding: A study of Hong Kong. Architectural Science Review. 2011;**54**:305-315

[50] Unger J. Connection between urban heat island and sky view factor approximated by a software tool on a 3D urban database. International Journal of Environment and Pollution. 2009;**36**:59-80

[51] US Green Building Council. LEED Reference Guide for Neighborhood Development. Version 4. Washington DC: U.S. Green Building Council; 2021

[52] Civic Exchange Hong Kong. Open Space Handbook: A guide for journalists in Hong Kong. Hong Kong: Civic Exchange; 2018

[53] Home R, Bauer N, Hunziker M. Cultural and biological determinants in the evaluation of urban green spaces. Environment and Behavior. 2010;**42**(4):494-523

[54] Fung W. Heat Island study—Satellite derived air temperature. Journal of Applied Meteorology and Climatology. 2009;**48**(4):863-872

[55] HK Green Building Council. BEAM Plus New Buildings, Version 2.0. Hong Kong; 2021

[56] Salim M, Schlünzen H, Grawe D. Including trees in the numerical simulations of the wind flow in urban areas: Should we care? Journal of Wind Engineering and Industrial Aerodynamics. 2015;**144**:84-95

[57] Giometto MG, Christen A, Egli PE, Schmid MF, Tooke RT, Coops NC, et al. Effects of trees on mean wind, turbulence and momentum exchange within and above a real urban environment. Advances in Water Resources. 2017;**106**:154-168

[58] Park M, Hagishima A, Tanimoto J, Narita K. Effect of urban vegetation on outdoor thermal environment: Field measurement at a scale model site. Building and Environment. 2012;**56**:38-46

[59] Kong L, Lau KKL, Yuan C, Chen Y, Xu Y, Ren C, et al. Regulation of outdoor thermal comfort by trees in Hong Kong. Sustainable Cities and Society. 2017;**31**:12-25

[60] Ng E, Cheng V. Urban human thermal comfort in hot and humid Hong Kong. Energy and Buildings. 2012;**55**:51-65

[61] Development Bureau HKSAR. Public Open Space in Private Developments Design and Management Guidelines. Hong Kong

[62] Architectural Services Department. Elderly-friendly Design Guidelines [Internet]. Hong Kong. 2022. Available from: https://www.archsd.gov.hk/media/reports/practices-and-guidelines/20190326_5501_Elderly-friendly%20Design%20Guidelines_FINAL.pdf

[63] Li Y, Song Y. Optimization of vegetation arrangement to improve microclimate and thermal comfort in an urban park. International Review for Spatial Planning and Sustainable Development. 2019;**7**(1):18-30

[64] Jo H. Impacts of urban greenspace on offsetting carbon emissions for middle Korea. Journal of Environmental Management. 2002;**64**:115-126

[65] Yin S, Shen Z, Zhou P, Zou X, Che S, Wang W. Quantifying air pollution attenuation within urban parks: An experimental approach in Shanghai, China. Environmental Pollution. 2011;**159**(8):2155-2163

[66] Beckett K, Freer-Smith P, Taylor G. The capture of particulate pollution by trees at five contrasting urban sites. Arboricultural Journal. 2000;**24**:209-230

[67] Jim C, Chen W. Assessing the ecosystem service of air pollutant removal by urban trees in Guangzhou (China). Journal of Environmental Management. 2008;**88**(4):665-676

[68] International WELL Building Institute pbc. The WELL Building Standard V1 with Q1 2020 Addenda. USA; 2020

[69] James Hutton Institute, University of Edinburgh, University of Glasgow, Heriot Watt University, Biomathematics and Statistics Scotland. Green space Services: Community Engagement Case Study—Environment and Human Health

[70] Francis M. How Cities Use Parks for Community Engagement: A Guide for Mayors. In: City Parks Forum. American Planning Association; 2002

Section 2

The Upcoming Trends in Urban Green Spaces

Chapter 6

Green Stormwater Infrastructure in the Context of the European Green Deal Policy

Başak Kılıç Taşeli

Abstract

It is thought that this section will make an important contribution to the literature and at least reflect the strong link between green stormwater infrastructure and the European Green Deal Policy to the readers. The European Green Deal has targets covering many different sectors, including construction, biodiversity, energy, transportation, and food, which include the enactment of new laws on green rainwater infrastructure. Green stormwater infrastructure not only controls stormwater volume and timing but also supports the benefits ecosystems bring to us. Stormwater is defined as rainwater or melted snow runoff from streets, lawns, and other areas. When rainstorm water is absorbed into the soil, it is filtered and eventually replenishes aquifers or flows into streams and rivers. Runoff carries sediments, nutrients, or other pollutants into water sources that degrade water quality, threaten drinking water supplies, and complicate water treatment processes. When drought concentrates pollutants, it can further limit dilution, making worse conditions. In order to prevent the problems caused by inefficient rainwater management systems, green infrastructure applications that mimic natural habitats, absorb excess water, and help protect water while preserving water quality have gained importance in recent years.

Keywords: the European Green Deal, green infrastructure, stormwater management, rainwater, sustainability

1. Introduction

The European Green Deal aims to make the European Union climate neutral by 2050 and reduce greenhouse gas emissions by at least 50% and 55% compared to 1990 levels, transforming the EU into a resource-efficient competitive economy and leaving no one and nowhere behind signed for. This agreement will require a review of every single law on climate, circular economy, building renovation, biodiversity, farming, and innovation. It also has targets covering many different sectors, including construction, biodiversity, energy, transportation, and food. **Figure 1** shows the various elements of the European Green Deal.

The most striking result of this strategy is that it includes potential carbon tariffs for countries that do not reduce their greenhouse gas emissions at the same rate.

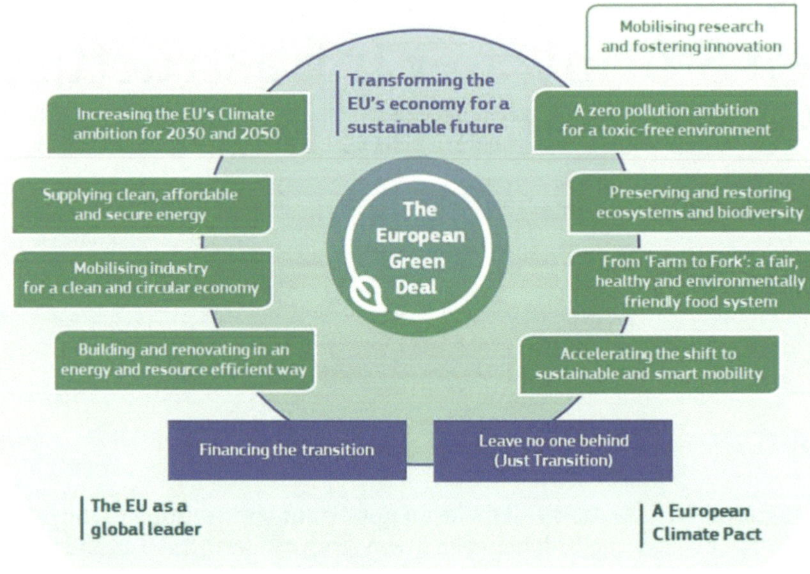

Figure 1.
The various elements of the European Green Deal [1].

The mechanism to achieve this is called the Carbon Limit Adjustment Mechanism (CBAM). Other mechanisms of this strategy are circular economy action plan, Emissions Trading System (ETS), Farm to Fork strategy, possible revision of all climate-related policy instruments, sustainable and smart mobility strategy, and forest protection and restoration. As the first step, Member States must update their climate and national energy plans in 2023 to match the EU's 2030 climate target.

The European Commission has announced the EU Energy System Integration Strategy for cleaner and greener energy. This strategy includes measures to develop clean fuels (including hydrogen) to develop a circular system.

Another strategy to achieve the EU's climate targets is the Circular Economy Industrial Policy. The EU has announced this strategy in order to revitalize industrial zones and have the best technologies. Circular economy-friendly commodity markets are a strategy that requires the decarbonization and modernization of energy-intensive industries such as steel and cement. The "sustainable products" policy, which focuses on reducing waste of materials (textile, construction, vehicle, battery, electronics, and plastic) and aims to reuse products and strengthen recycling processes, is also envisaged.

The building and renovation policy focuses on promoting the use of energy-efficient building methods, increasing digitalization, and tripling the renovation rate of all buildings.

The Commission's Zero Pollution Action Plan aims to ensure that no pollution comes from all industrial activities by 2050. This strategy also includes the substitution of toxic substances used in industries with toxic ones.

Reducing emissions from transport methods is another target area under the European Green Deal. It includes the implementation of a comprehensive strategy on "Sustainable and Smart Mobility" that will increase the use of sustainable and alternative fuels in land, sea, and air transport, set emission standards for internal combustion engine vehicles, and offer a sustainable alternative. As a solution, the development of smart traffic management systems and applications, smart load distribution management, and the creation of charging points for electric vehicles are among the objectives of this strategy.

Sustainable finance, or green finance, is supported by the adoption of the Paris Climate Agreement, which requires parties to build financial flows consistent with low greenhouse gas emissions and a pathway to climate-resilient development.

The Farm to Fork' strategy is a strategy that includes the support given to farmers and fishermen as well as the production and transfer methods of these resources, in addition to the issue of food sustainability, and adopts a climate-friendly approach. Specific target areas include reducing chemical pesticide use, increasing the availability of healthy food options, and helping consumers understand the health ratings of products and sustainable packaging. The Farm to Fork strategy is geared toward a better balance of nature, food systems, and biodiversity and includes the following objectives: [2].

- To organicize 25% of EU agriculture by 2030.
- To reduce pesticide use by 50% by 2030.
- To reduce the use of fertilizers by 20% by 2030.
- To reduce nutrient loss by at least 50%.
- To reduce the use of antimicrobials in agriculture and antimicrobials in aquaculture by 50% by 2030.
- To establish sustainable food labeling.
- To reduce food waste by 50% by 2030.

The biodiversity strategy is an important part of the EU's climate change mitigation strategy and includes the allocation of 25% of the European budget to combating climate change in order to restore biodiversity. The EU Biodiversity Strategy for 2030 includes the following objectives: [2].

- Conservation of 30% of sea lands and 30% of land lands, especially primary forests and old forests.
- Planting 3 billion trees by 2030.
- Restoring at least 25,000 kilometers of rivers for free flowing.
- 50% reduction in pesticide use by 2030.
- Increasing organic agriculture.
- Increasing biodiversity in agriculture.
- Reversing the decline of pollinators.

The development of green infrastructure is an important step toward the success of the EU 2020 Biodiversity Strategy, with the strategy's second objective requiring "by 2020, the maintenance and development of ecosystems and their services by establishing green infrastructure and restoring at least 15% of degraded ecosystems." However, green infrastructure contributes to the full implementation of all six objectives of the strategy – specifically the Birds and Habitats Directive (target 1) and to the conservation and enhancement of biodiversity in the wider rural and marine environment (targets 3 and 4) [3].

The Commission has adopted an EU-wide strategy that encourages investment in green infrastructure to restore the health of ecosystems, keep natural areas connected, allow species to thrive in their natural habitat, and promote green infrastructure across Europe. It is designed to provide ecosystem services such as green infrastructure planning, water purification, air quality, recreation space, and climate mitigation, providing environmental, economic, and social benefits through natural and seminatural solutions. This network of green (land) and blue (water) areas improves environmental conditions, thereby increasing the health and quality of life of citizens, creating job opportunities, increasing biodiversity, and supporting the green economy [4].

2. Green stormwater infrastructure

As it is known, the storm or flow moves over the ground by gravity and reaches streams, rivers, ponds, lakes, and oceans. With urbanization, impermeable surfaces (roofs, streets, parking lots, and sidewalks) increase, so the permeable areas decrease and as a result, the volume of storm water flow increases. Stormwater management strategies include the prevention, containment, and reuse of stormwater runoff.

Green stormwater infrastructure is systems like biological ditches, catchments, wetlands, rain gardens, and urban green spaces that mimic natural processes that result in infiltration, evaporation, or use of rainwater. It is becoming increasingly clear that green stormwater infrastructure not only controls stormwater volume and timing but also supports ecosystem services, which are the benefits ecosystems provide to humans.

"Green Infrastructure can be broadly defined as a strategically planned network of high-quality natural and seminatural areas with other environmental features, which is designed and managed to deliver a wide range of ecosystem services and protect biodiversity in both rural and urban settings. More specifically GI, being a spatial structure providing benefits from nature to people, aims to enhance nature's ability to deliver multiple valuable ecosystem goods and services, such as clean air or water" [5].

"Green infrastructure is a strategically planned network of natural and seminatural areas with other environmental features designed and managed to deliver a wide range of ecosystem services. It incorporates green spaces (or blue if aquatic ecosystems are concerned) and other physical features in terrestrial (including coastal) and marine areas. On land, GI is present in rural and urban settings" [6].

" Green infrastructure is an approach to wet weather management that uses soils and vegetation to utilize, enhance, and/or mimic the natural hydrological cycle processes of infiltration, evapotranspiration, and reuse" [7].

Traditional stormwater infrastructure, or gray infrastructure, is designed to divert rainwater from urban areas through pipes and channels. Runoff from these

DOI: http://dx.doi.org/10.5772/intechopen.103917

Figure 2.
Potential components of a green infrastructure [9].

surfaces can suffocate infrastructure systems, carrying pollutants into streams, rivers, and lakes when rainwater runs off impermeable streets, parking lots, sidewalks, and roofs. Flows can also cause erosion and flooding that can damage property, infrastructure, and wildlife habitat. In addition to flow problems, impermeable surfaces also prevent water from penetrating the soil and refilling groundwater sources.

Green infrastructure uses plants, soils, landscape design, and engineering techniques to contain, absorb, and reduce polluted stormwater runoff. Green infrastructure prevents combined sewer overflows for cities by preventing or reducing the amount of runoff flowing directly into storm drains. It provides many environmental, social, and economic benefits such as improved surface water quality, water savings, and improved esthetics.

Improved water quality, increased water supply, reduced flooding, improved air quality, greater resilience to climate change, and increased habitat improvement are examples of environmental benefits of GI. The economic benefits are increased property values and reduced filtering costs and infrastructure cost savings. Choosing green infrastructure over traditional gray infrastructure such as concrete pipes has the following advantages [8]:

- It is more attractive, effective, and multifunctional.
- Provides habitat for beneficial wildlife.
- It can eliminate mosquito breeding habitat.
- Infrastructure costs are reduced compared to gray infrastructure.
- Eliminates stagnant water that could serve as a mosquito breeding habitat.

Restoring floodplain forest instead of building a dam not only benefits the local community but also forests with a good mix of species and ages can absorb large amounts of water, protecting the soil and preventing or at least reducing the effects of floods and landslides. It also provides habitats for animals, offers recreation areas, and contributes to reducing global climate change through carbon sequestration. Properly designed parks, city gardens, green roofs, and walls all contribute to biodiversity. In addition, it helps combat climate change, significantly improves the health and well-being of urban residents, and improves social cohesion and the quality of the living environment. **Figure 2** represents potential components of green infrastructure.

3. Green stormwater infrastructure techniques

Green infrastructure techniques use soils and vegetation to infiltrate, evaporate, and/or recycle stormwater runoff and techniques for sustainable rainwater management are as follows [10]:

- Green roofs or eco-roofs.
- Downspout disconnection.
- Rainwater harvesting and reuse systems (water retention tanks).
- Natural stormwater drainage systems (bioswales and vegetative swales).
- Stormwater planters, curb extensions, and infiltration gardens.
- Stormwater basins (detention and retention basins, and ponds).
- Efficient landscaping and irrigation.
- Reduce the area of impervious surfaces and use of pervious paving (pervious asphalt, concrete and structural grass-paving systems).
- Stream rehabilitation and daylighting projects.

3.1 Urban forests

Urban forests, as the name suggests, are an important component of urban green infrastructure systems in cities. These forests are esthetically valuable forests that reduce the need for maintenance and irrigation and use suitable tree and plant species instead of harmful and invasive species.

The benefits of urban forests [1] are as follows:

- Energy savings: It has been found that trees planted around houses can save up to 47% of energy for heating and cooling.
- Urban heat island: Thanks to the evaporative cooling mechanism, wooded areas have been found to have lower air temperatures than open areas without trees. Urban heat islands negatively affect human health by causing heat to intensify due to the materials and infrastructure used.
- Water management: It has been found to help city water management by diverting large amounts of precipitation hitting trees through water channels through urban forests.
- Air pollution: Trees act as carbon sinks, improving air quality in cities.
- Property values: It has been found that the value of the property increases as the amount of trees above its value increases.
- Public health: It shows that urban greenery can positively improve people's mental and mental health.

3.2 Constructed wetlands

Wetlands are often built on floodplains, and with proper planning and management, they help prevent changes in natural hydrology and the introduction of invasive species. By using the natural processes of plants, soils, and microorganisms, it can hold a large part of suspended solids, slow down the water flow, and provide the purification of pollutants.

The benefits of constructed wetlands are as follows [1]:

- Water efficiency: Constructed wetlands attempt to mimic natural wetland ecosystems. They were built to improve water efficiency and water quality.
- Cost-effective: Wetlands have low operating and maintenance costs.
- Esthetics: Built wetlands can add greenery to their surroundings.

3.3 Green and blue roofs

Green roofs improve air and water quality while reducing energy costs. Plants and soil provide more green space and insulation on roofs. Green and blue roofs help reduce urban runoff by trapping precipitation in stormwater management in dense urban areas. Green roofs also act as a carbon sink, trapping carbon pollution. It has been found that 40–80% of the total rain volume falling on green roofs can be reserved. Since the water released from the roofs will flow at a slow rate, they cause a decrease in the amount of runoff entering the basin at once. Although blue roofs do not technically count as green infrastructure, they function to collect and store precipitation, reducing the flow of water flowing into sewer systems. Blue roofs also help to save energy by reducing cooling costs.

Other benefits are as follows [11]:

- Reducing ambient air temperatures by shading the roof in summer.
- Prevention of heat transfer in winter.
- Absorption, retention, filtering, and storage of precipitation.
- Minimizing the effects of impermeable surfaces.
- Extending the life of roofing membranes.
- Increasing green space and wildlife habitat.
- Developing city esthetics.
- Improving air quality by acting as a carbon sink.
- Reducing runoff and benefiting the capacity of urban stormwater infrastructure.

Some advantages of green roofs over reflective roofs [11] are as follows:

- They use radiant energy to cool the ambient air instead of reflecting sunlight onto adjacent buildings.
- They can resist UV radiation.
- They can ensure a long life by protecting the roof membrane.
- They increase the esthetic value by providing green space.

Some disadvantages of green roofs over reflective roofs [11] are as follows:

- Higher initial costs due to additional material and higher installation costs.
- Some ongoing maintenance is also required, such as fertilizing the vegetation annually and checking the plants until the plants are fully established and the roof is fully covered.
- Since they are not mechanically fixed to the roof, they are sensitive to strong winds.
- Any leak in the waterproofing material is difficult to find.
- They do not work well on very steeply pitched roofs.

3.4 Rain gardens

Rain gardens are areas with shrubs and plants that collect rainwater from roofs or sidewalks, allowing the rainwater to seep slowly into the ground. It is an alternative to reduce urban water flow and slow the flow of water by infiltration. In residential applications, water flow can be reduced by 30% by using rain gardens in the garden of the house.

3.5 Downspout disconnection

Downspout disconnection is a form of green infrastructure that separates roof downspouts from the sewer system and directs roof water flow to permeable surfaces, used to store rainwater or allow water to penetrate the ground. It is clear that disconnecting the downspout would be particularly beneficial in cities with combined sewer systems.

3.6 Bioswales

Bioswales, much like rain gardens, can be depicted as channels with vegetation or much often placed in long narrow spaces in urban areas. They absorb runoff or carry stormwater runoff from heavy rains into sewers or directly into surface waters. Vegetative bioswales infiltrate, slow, and filter stormwater streams that are most beneficial along streets and parking lots.

3.7 Green alleys

It can be characterized as redesigning existing passages to reflect more light to reduce the heat island effect, capture rainwater, and make the space usable. In addition to its ecological benefits, green alleys provide open spaces by transforming previously unsafe areas into playgrounds and walking/bike corridors.

4. Conclusion

Combined sewer systems that include sewage and stormwater streams are unfortunately common in many older municipalities. Urban sustainability, on the other hand, is the management of stormwater runoff, which has become an increasingly challenging issue with the increasing amount of impermeable space in urban environments. Increasing political, social, and environmental pressures are forcing engineers to reduce the number and severity of floods or combined sewer overflows.

Summing up the benefits of green infrastructure for people living in cities, it causes a drastic increase in biodiversity that supports rich wildlife communities, including many rare and endangered species [12, 13], its significant impact on climate mitigation due to the reduction of urban heat islands [14, 15], its positive effects on the control of runoff [16, 17], and removing air-borne pollutants [18].

EU strategy on green infrastructure refers to the goal of "no net loss of biodiversity and ecosystem services by 2020." This approach should include no net loss of green infrastructure. This policy incorporating the concepts of water, biodiversity, soil, climate, air, and landscape thus allowing green infrastructure is to be addressed in a holistic issue.

This chapter intends to provide the reader with a comprehensive overview of urban and rural area stormwater management, green infrastructure practices, green stormwater solutions, benefits of green stormwater infrastructures, and climate adaptation to stormwater runoff.

The author advocates the concept of green infrastructure, claiming to offer a progressive planning approach that facilitates synergies between economic growth, environmental protection, and social development.

Author details

Başak Kılıç Taşeli
Faculty of Engineering, Environmental Engineering Department, Giresun University, Giresun, Turkey

*Address all correspondence to: basak.taseli@giresun.edu.tr

References

[1] Available from: https://eur-lex.europa.eu/legal-content/EN/TXT/HTML/?uri=CELEX:52019DC0640&from=ET

[2] Available from: https://en.wikipedia.org/wiki?curid=62693289 [Accessed: December 30, 2021]

[3] Available from: https://ec.europa.eu/environment/pubs/pdf/factsheets/biodiversity_2020/2020%20 Biodiversity%20Factsheet_EN.pdf [Accessed: December 29, 2021]

[4] Available from: https://ec.europa.eu/environment/nature/ecosystems/strategy/indexen.htm [Accessed: December 29, 2021]

[5] European Union: Building a Green Infrastructure for Europe. 2013. Available from: https://zrsvn-varstvonlarave.si/wp-content/uploads/2019/10/Trbizan-Kranjc-Vertej_PERFECT.pdf [Accessed: December 30, 2021]

[6] German Federal Agency for Nature Conservation (BfN). Federal Green Infrastructure Concept. 2017. Available from: https://www.interreg-central.eu/Content.Node/Definitions.html [Accessed: December 28, 2021]

[7] US Environmental Protection Agency. Managing Wet Weather with Green Infrastructure: Action Strategy. 2008. Available from https://www3.epa.gov/npdes/pubs/gi_newsletter_apr08.pdf. [Accessed: December 29, 2021]

[8] Available from: https://www.epa.gov/sites/default/files/2016-10/documents/green_infrastructure_brochure_final.pdf [Accessed: December 26, 2021]

[9] Available from: https://biodiversity.europa.eu/topics/green-infrastructure [Accessed: December 25, 2021]

[10] Available from: https://www.epa.gov/sites/default/files/2016-10/documents/green_infrastructure_brochure_final.pdf. [Accessed: December 22, 2021]

[11] DOE/EE-0298. Available from: www.eere.energy.gov/femp/Federal_technology_alert [Accessed: December 23, 2021]

[12] Gilbert OL. The Ecology of Urban Habitats. Dordrecht, The Netherlands: Springer; 1991

[13] Zhang H, Lu S, Fan X, Wu J, Jiang Y, Ren, Wu J, Zhao H. Is sustainable extensive green roof realizable without irrigation in a temperate monsoonal climate? A case study in Beijing. Science of The Total Environment. 2021;**753**: 1-7. Article: 142067. DOI: 10.1016/j.scitotenv.2020.142067

[14] Payne S, Barker A. Implementing green infrastructure through residential development in the UK. In: Handbook on Green Infrastructure. Cheltenham, UK: Edward Elgar Publishing; 2015. pp. 375-394

[15] Smalls-Mantey L, Montalto F. The seasonal microclimate trends of a large scale extensive green roof. Building and Environment. 2021;**107792**. DOI: 10.1016/j.buildenv.2021.107792-107802

[16] Mentens J, Raes D, Hermy M. Green roofs as a tool for solving the rainwater runoff problem in the urbanized 21st century? Landscape and Urban Planning. 2006;**77**:217-226

[17] Armson D, Stringer P, Ennos AR. The effect of street trees and amenity grass on urban surface water runoff

in Manchester, UK. Urban Forestry & Urban Greening. 2013;**12**:282-286

[18] Shafique M, Xue X, Luo X. An overview of carbon sequestration of green roofs in urban areas. Urban Forestry & Urban Greening. 2020;**47**:126515-126529

Chapter 7

Nature-Based Solutions for Climate Change Adaptation in Scholar and Social Buildings: Concepts and Pilot Solutions for Portugal and Spain

Teresa Batista, Salustiano Torre, José Fermoso, Nuno Morais, Ramona Avila, Carlos Pinto Gomes, Rui Alexandre Castanho and José Cabezas Fernández

Abstract

Climate change has been recognized as one of the most serious environmental, social, and economic challenges facing the world today. Contextually, the Inter-municipal Plan for Climate Change Adaptation in Alentejo Central (PIAAC-AC) has already identified the tendencies and future scenarios of climate change in Alentejo Central until the end of the twenty-first century, namely the increase in the number of days with very high temperatures, the number of tropical nights and heat waves, and the general decrease in annual rainfall. In this scenario, the concerns with school communities and users of social services increase. The project "LIFE-myBUILDINGisGREEN"—"application of nature-based solutions for local adaptation of educational and social buildings to Climate Change," developed in partnership with CIMAC (Portugal), CARTIF Technology Center (Spain), Diputación de Badajoz and CSIC—Consejo Superior de Investigaciones Cientificas (Real Jardin Botanico— Spain—Project Leader), and the Porto City Council (Portugal), focuses on the construction sector, in particular on education and social services buildings in cities in Europe. It aims to implement the prototypes (building adaptation) of nature-based solutions (NBS) on walls, roofs, playgrounds, and exterior surfaces on three pilot buildings. The overall objective is to contribute to improve resilience in these buildings using autochthone vegetation.

Keywords: climate change adaptation, nature-based solutions, green roofs, green walls, educational buildings

1. Introduction

Climate change has been recognized as one of the most serious environmental, social, and economic challenges facing the world today. Several studies in Portugal and Spain recognize the tendencies for climate change at regional level [1–6].

In Alentejo Central, the Intermunicipal Plan for Climate Change Adaptation (PIAAC-AC) [1] has already identified the tendencies and future scenarios of climate change in Alentejo Central until the end of the XXI century, namely the increase in the number of days with very high temperatures, the number of tropical nights and heat waves, and the general decrease in annual rainfall. In this scenario, the concerns with school communities and users of social action services increase.

The project "LIFE-myBUILDINGisGREEN"—"application of nature-based solutions for local adaptation of educational and social buildings to climate change," developed by CIMAC in partnership with the CARTIF Technology Center (Spain), Diputación de Badajoz, the CSIC—Consejo Superior de Investigaciones Cientificas (Real Jardin Botanico—Spain), and the Porto City Council, focuses on the construction sector, in particular on education and social services buildings in all cities and towns in Europe. It aims the implementation of 3 prototypes (building adaptation) of nature-based solutions (NBS) on walls, roofs, exterior surfaces, and car parking of 3 pilot buildings, one located in Évora—Alentejo Central—South of Portugal, one in the city of Porto in the north of Portugal and one in Solana de los Barros in the Badajoz Province—southwest of Spain. The overall objective is to contribute to improve the resilience in these buildings using NBS and autochthone vegetation. This paper presents the LIFE-MyBuildingisGreen project, its objectives, and expected results [7].

Climate change is now recognized as one of the most serious environmental, social, and economic challenges facing the world. The IPCC's Fifth Report (2014) [8] identifies that many of the global risks of climate change are concentrated in urban areas. In addition, implementing measures that enhance resilience and enable sustainable development can accelerate processes of adaptation to climate change [9].

Europe's educational and social service buildings will face multiple challenges in the coming decades, and climate change will add pressure to it. The project focuses on the building sector, specifically on the public buildings dedicated to education and social services existing in all cities and towns in Europe. The impacts of climate change (heat waves and changes in annual and seasonal precipitation patterns) are affecting the health and well-being of children and elderly people who are the main users of these types of buildings [10–12].

The overall objective of the project is to contribute to increasing the resilience of these buildings by implementing in them NBS as prototypes of climate adaptation and improved well-being. Additionally, other specific objectives are:

1. Improve knowledge base on the development, assessment, and monitoring of vulnerability to climate change of buildings through developing and testing a common, ready-to-use method in the southern, western, central, and northern European regions.

2. Analyze and verify the impact of NBS as adaptation measures to climate change in three pilot buildings.

3. Promote sustainable ecosystem-based adaptation measures and increase the capacity to apply such knowledge in practice through dissemination and demonstration workshops.

4. Promote the governance of regional authorities, directors and building managers, and the building industry to integrate NBS as part of measures to promote

sustainable adaptation in buildings through the development, creation, and dissemination of governance tools and territorial agreements.

5. Contribute to the development and implementation of a common policy at the UE level on adaptation and climate change by transferring best practices and knowledge base to target groups and stakeholders.

The main actions of the project are:

- Selection of 3 pilot buildings by weighting the technical criteria that characterize each building. Analysis of European databases with information relevant to the implementation of NBS in public buildings of education and social services.

- Evaluation and analysis of the baseline of each building to determine the initial state of the building and to use it as a reference to assess the impact of the NBS using the indicators proposed within the Project.

- Installation and implementation of 3 prototypes based on nature in facades and partition walls, roofs, and exterior surfaces of each selected pilot building. In addition, it will be complemented by the implementation of sustainable measures of induced natural ventilation, seasonal shading (natural and artificial), and the choice of native Mediterranean and Atlantic species.

- Monitoring of the impact of the works carried out with the aim of verifying and evaluating the importance of these measures as alternative solutions to climate adaptation for regional authorities at the local level.

- Creation of actions and tools of governance that make possible the signing of government agreements, the development of inter-municipal programs, and the integration of the NBS into regulations and multi-year action plans of political, technical, and institutional scope.

- Demonstrating the feasibility and transferability of NBS as measures for sustainable adaptation to stakeholders in policy, building, climate change, building, and urban planning at local, national, and European levels. Creation of capacities to improve knowledge about the NBS to the responsible authorities and users, directors, and managers of education and social services centers and the building sector.

- Communication and dissemination of the results and good practices of the Project through the website, bulletin board, newsletters, brochures, press articles, and networking at regional, national, and European levels and dissemination activities online with all interest groups and stakeholders.

In the case of Portugal and specifically in Alentejo and Oporto (areas of action of the Project in Portugal), the residential park in 2011 was around 111,826 accommodations. There are more than 120 public buildings of responsibility and municipal management that serve around 238,000 people. Of these, 49 correspond to public elementary schools and the others refer to administrative, institutional, cultural, and sports buildings, scattered throughout Alentejo and Oporto. These buildings house more than 3000 employees. In Portugal, it is the state that assumes the ownership of the buildings of education and social services, but the management and maintenance are carried out in a shared way between the state and the municipalities. The vulnerability to climate change of these buildings, located in regions and municipalities of Spain and Portugal, is a consequence of the high temperatures that are reached inside them between the months of May to October and the constructive characteristics of these buildings [13].

Part of the heating problem of these buildings can be attributed to the construction typology and the materials used. In Spain and Portugal, many of the education and social services centers were built prior to the basic thermal conditions' regulation (Royal Decree 2429/79, of July 6, which approves the Basic Building Standard NBE- CT-79, on Thermal Conditions in buildings). This document provided the basic design criteria that allowed damping of the thermal wave through the enclosures. It was from that moment when thermal insulation incorporated into double-leaf walls began to be used, achieving a notable improvement in the thermal performance of the facades with respect to the mass walls used in previous years. This regulation was replaced by the current Technical Building Code, in 2006, in which special attention has been given to reducing energy demand with the incorporation of high-performance solutions in the enclosures.

As a result, we find insufficiently thermally insulated buildings, and with little sun protection in window openings, which in hot climates such as those in Spain or Portugal, produces overheating effects in last spring, summer, and early autumn periods [14].

In addition, regulatory and labor regulations establish that the temperature inside these buildings should not be less than 10°C or higher than 27° since it allows risk due to thermal stress and causes health problems for the workers. The company or public authority responsible must take the necessary preventive measures and the work and teaching activity should be paralyzed if the thermal risk is maintained.

In this regard, it is worth mentioning that there is currently a significant climate problem and social impact in the regions of Southern Europe, specifically in Spain and Portugal.

This problem is related to the high temperatures inside education and social services centers that users suffer regularly between May and October, with the decrease in environmental comfort in these buildings and with the worsening conditions of health and well-being of the users. So much so that the aforementioned climate problem is addressed by the digital and written press of countries such as Spain, Portugal, Italy, and France that echo on the front pages of this pressing situation in educational and social centers.

The solution to this problem cannot go through the application of a Climate Control Program that allows the installation of artificial cooling by means of air conditioners in these public centers, since neither the responsible public authorities nor the European guidelines advise it, because they contribute to global warming, due to its high consumption of electrical energy and emission of CO_2 into the atmosphere [15]. In addition, air conditioners have contributed to increasing the percentage of citizens with respiratory problems such as laryngitis or pharyngitis and have exacerbated the chronic processes of the population at risk. To date, the solutions that the administrations responsible for these centers are taking are the regularization of the students teaching hours, allowing parents to take their children ahead of time as a result of the maximum peaks of indoor temperature coinciding with the hours Central of the day. This situation is causing disorganization of schedules that makes it more difficult to reconcile work and family life [11].

In the sense of the framework of a local adaptation strategy against the effects of climate change in these buildings, it must generate governance tools between the different regional and local administrations with responsibilities in the field of education and social welfare, which allows these centers to innovate and rehabilitate by implementing nature-based solutions that allow to respond to periods of potential overheating and minimize unwanted heat gains during the day and night. The application of these NBS tools in public education and social services centers will enable responsible local administrations to fulfill the objective of the Covenant of Mayors for climate and energy, as well as offer natural, and sustainable response related to the environmental comfort and temperature inside these buildings.

2. Climate tendencies and prospective scenarios in study areas

2.1 Climatic situation of the territory

Alentejo's central climate is typical Mediterranean. It is a mesothermal climate type with rainy winter and hot, dry summer. However, the geographical position of the Alentejo Central and the layout of the main relief masses of southern Portugal give its (Mediterranean) climate a certain continental character [1].

In general, Alentejo Central has an air temperature regime, with average annual temperature values around 16°C, with a relatively high annual temperature range in the national context, which is accentuated inland. Winter is cool, with average temperatures slightly below 10°C and average minimum temperatures around 5°C, with January being the coldest month.

The summer in the Alentejo Central is hot or very hot, with the heat conditions inland, as well as in the most sheltered and/or less ventilated places. In this context, average temperatures in the warmer months (July and August) range from 23 to 25°C, with average maximum temperature values exceeding 30°C.

The most significant trends observed from 1971 until 2015 were an increase in the mean annual temperature, with a greater incidence in spring and summer maximums, with a greater frequency of tropical nights. In addition, there has been a reduction in precipitation in summer and autumn and an increase, in this last season, in precipitation events concentrated in short periods.

In general, the climate of Badajoz can be described as the Mediterranean with slight Atlantic-continental nuances. Temperatures are very high in summer and mild in winter, with a substantial incidence of sunlight and long hours of sunshine, with the maximum values in the westernmost part, near Portugal. An essential thermal amplitude stands out, evidencing the marked seasonality typical of the Mediterranean climate. Precipitation is the primary source of humidity for the region. Although the general precipitation balance is negative from all points of view since it is scarce in most of the territory, there is also significant evaporation due to the high temperatures, making these rains even more insufficient [16].

As far as rainfall behavior throughout the year is concerned, in Badajoz, there is an evident dry season, summer. There is rainfall in the rest of the year, although with greater abundance toward the end of autumn and beginning of winter, with less critical rains in the rest of autumn and during spring.

In recent decades in the territory of Badajoz, there has been a tendency to increase the minimum temperature while spring precipitation decreases and autumn precipitation increases. An increase in the number of hot days is occurring, as is a decrease in the number of cold days.

The city of Porto is part of the NW of the Iberian Peninsula, in the coastal strip of the Atlantic Ocean, suffering the influence of the sea currents that dominate the North Atlantic, which find in this area of the Iberian Peninsula the first contact with a continental area. Despite belonging to the Atlantic climatic subtype, the climate in Porto has particular characteristics as a result of the influence and attributes, and positioning of an important barometric apparatus—the Anticyclone of the Azores.

In the last 40 years, there has been a steady pace characterized by a moderately cool winter in the areas closest to the sea and cold or very cold in the most sheltered areas of the interior and at the highest altitudes, a moderately warm summer often influenced by the winds of NW and the morning advection fogs along the coast, and hot or very hot in areas far from the Atlantic moderating action or higher altitude [6]. Precipitation is more frequent and intense in the winter months but can occur throughout the year. Winters in O'Porto have tended to be moderately cool along the coast and cold to very cold inland. While the summers have been moderately warm by the sea, they become hot inland. Precipitation tends to be concentrated toward the end of autumn and the beginning of winter.

2.2 Climate scenarios

Climate scenarios are made through the collection and treatment of future climate information (projections) using different models and for different global climate scenarios, serving as support to identify possible changes in future climate.

As a result of the global problem of climate change, both in Portugal and Spain, studies have been developed to analyze the scenarios that could occur in the future [1–5]. In the different scenarios, the most extreme weather conditions will be reinforced, generating critical variations in the meteorological variables. Temperatures will suffer a general increase in the annual average in the areas studied. Very hot days will increase in summer, autumn, and spring. Rainfall will suffer a decrease, accompanied by shifts in its distribution between the different seasons, with an increase in autumn.

Due to these scenarios and the actual conditions of the educational buildings referred in the introduction, the partnership of the project "MybuildingisGreen" [7] decided to study and implement in each one of the selected buildings the solutions as explained in the next section for the Évora pilot building.

3. Nature-based solutions (NBS): prototypes and pilot solutions in the study areas

In an increasingly urbanized world, nature—rich in intelligent and efficient solutions to the challenges of cities—has been forgotten in the design and management of urban spaces. Many of the green (vegetation) and blue (water) surfaces have been transformed over time into gray (impermeable) surfaces, with serious implications for the quality of life of the inhabitants and increasing environmental risks.

Currently, some of this natural engineering has been trying to rescue some of this natural engineering for cities in order to ensure sustainable, cost-effective, multifunctional, and flexible solutions to various environmental challenges. This movement recognizes that it is more advantageous—ecologically and economically—to work alongside nature and nature than against it.

The European Union developed the EU Research and Innovation policy agenda on Nature-Based Solutions and Re-Naturing Cities [17, 18], which aims to position the EU as a leader in "Innovating with nature" for more sustainable and resilient societies. The EU defines nature-based solutions "as solutions that are inspired and supported by nature, which are cost-effective, simultaneously provide environmental, social, and economic benefits and help build resilience. Such solutions bring more and more diverse nature and natural features and processes into cities, landscapes, and seascapes, through locally adapted resource-efficient and systemic interventions."

The project My Building is Green aims to strengthen and support cities, and EU within the shift of the old urban paradigm based on gray surfaces, and works to contribute with specific and innovative nature-based solutions (more efficient, more sustainable, and greener) to the adaptation of buildings and improve the bioclimatic comfort and quality of life of users [12–14], which also act (each school), for example, as inspiration and a starting point for shift and adapt to climate change in cities.

Hereafter, some innovative solutions developed and projected within the project will be presented. However, it has to be mentioned that there are multiple NBSs that can be applicated for the adaptation of buildings to climate change, according to the type of construction to the climate and the vulnerability expected [7].

3.1 Prototype roofs mBiGWTray for application in Évora pilot

This system was created by the CARTIF Team in collaboration with SingularGreen and consists of a multilayer tray to maintain cover vegetation that is encapsulated with a white waterproof sheet to collect rainwater and reduce water loss [7]. The design of the system, including the selection of the appropriate plant species, is done so as not to require the installation of auxiliary irrigation. In the upper part of the encapsulation, there are some holes for planting plant species.

Schematically, the system would be a system of extensive vegetative cover pocketed to make it more resistant to rainfall shortages (**Figure 1**).

This system allows the installation of the trays on the roof directly occupying 50% of the surface with the vegetal part initially and the rest with a white surface that avoids the excessive capture of thermal energy. The installation is carried out using a checked system in which the planting surfaces alternate with the water collection surfaces.

In the joining area of each module, there are some holes that allow the entrance of water from the zone of collected water toward the zone with the vegetal system.

The bags have a drainage hole at the surface of the tray to allow the greatest amount of water to be stored avoiding the pooling of the substrate in which the roots of the plants are found.

Both the weight of the tray itself and its flat design mean that, in principle, an auxiliary roof anchor system is not necessary. However, if the prevailing winds in the area were very strong, the system could be weighed down using draining aggregate as partial filling of the holes in the tray.

The components of the System are:

- The encapsulation material can be made of white EPDM or PVC in order to get adequate durability. In principle, white PVC is selected as the first option by price mainly. The durability of the two materials is similar for this application, both with a 10-year warranty.

Figure 1.
Prototype mBiGWTray: Schema and implementation.

- The tray is composed of support with cavities to contain the planting substrate of the plants to be included. This is a MODI H50 ventilated pavement plate that is covered with a non-woven felt geotextile and inserted into the capsule. This tray allows structuring the planting of the vegetation, providing rigidity for handling the bagging, and creating spaces for water storage inside.

- Between the tray and the geotextile is the material that acts as a substrate that can be both sheep's wool and sphagnum or other similar materials. This layer allows transport by capillarity from the water storage area to the upper zone.

- Above the geotextile is a layer of substrate encapsulated in a light geotextile. In this area, it is where it contains the root part of the vegetation that is integrated into the system.

- The characteristics of the selected species are underwater requirement, creeping and surface roots, native species, and species indicated as beneficial for pollinators. A plantation of 16 plants per tray would be carried out. Initially, there would be bands covered with vegetation and bands without vegetation but over the years one could have a complete green cover.

On the other hand, the system is compatible with drip irrigation that can be integrated into the base structure. A more homogeneous vegetation maintenance would be achieved throughout the year. However, the initial design has been carried out so that the implementation of an irrigation system is not necessary and the previous tests that are going to be carried out are aimed in this direction.

3.2 Prototype facades mBiGToldo for creating vertical green surfaces in Évora

This system has been designed to create vertical surfaces with vegetation of very low thickness to create shading with a contribution of humidity to the environment. The system consists of an impermeable sheet on which a nonwoven felt is adhered and a semi-woven substrate is projected. Due to the low substrate thickness, hydroponic irrigation is integrated that is distributed by gravity across the surface of the substrate (**Figure 2**). In the lower zone, a channel for collecting excess irrigation is integrated and returned to the irrigation station.

Irrigation is done through a hydroponic irrigation station with adequate programming to cover the needs of the vertical garden at all times of the year and the environmental conditions that are to be generated. It is necessary to connect the irrigation station, to be installed under the level of the mBiGToldo, with all the gardens, both to provide the irrigation and to collect [7].

The components of the System are:

- Support frame: Structure with the appropriate dimensions made with the material with which the vertical wall structure that will support the prototypes of the facade is built.

- Waterproof support. It must be a sheet-shaped material resistant to punching and tearing to facilitate fixing. The currently recommended material is a PVC awning sheet but more sustainable materials are being sought.

- Root fixing sheet based on nonwoven felt or rock wool.

- Mix of substrate and compatible seeds that is applied by projection.

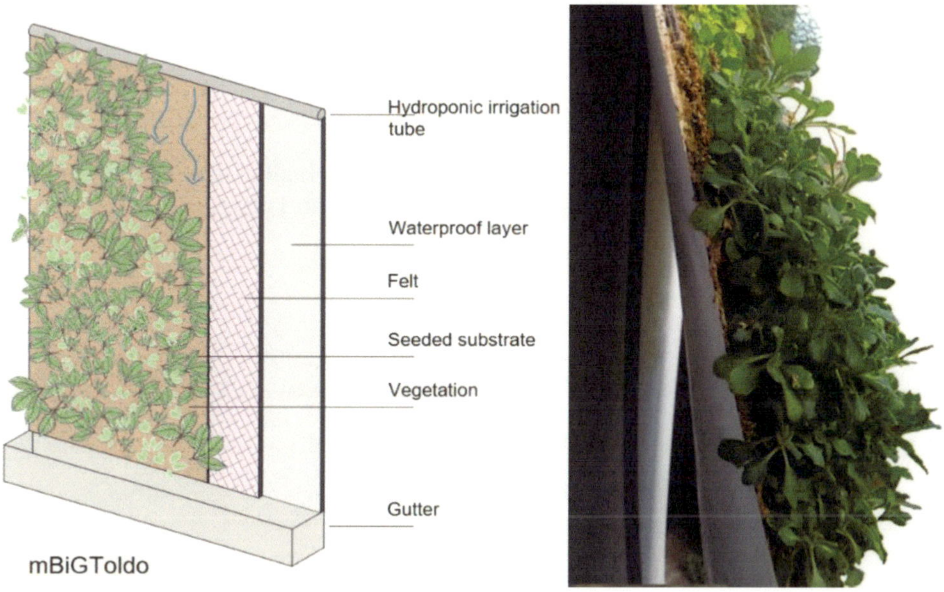

Figure 2.
mBiGToldo schema and implementation.

- The characteristics of the selected species will be easily propagated by seeds.
- Drip irrigation tube in the upper part and water collection gutter and injected to leftover water collection tube.

The system is only compatible with hydroponic drip irrigation that can be integrated into the support structure of the awning. The design includes a system for collecting excess irrigation water and returning it to the tank for optimization of water consumption.

Among the species that can be used are the *Festuca rubra*, *Agrostis stolonifera,* and *Sagina subulata*, but others can be evaluated depending on the location and suggestions made by the Royal Botanic Garden.

3.3 Prototype roofs mBi_GUL for application in Porto pilot

The mBi_GUL is a prototype solution inspired by Green Urban Living System (GUL), an innovative option develop under the Green Urban Living project (greenURBANLIVING), promoted by Amorim Isolamentos, Itecons, Neoturf, and ANQIP, and financed by European Commission through European Regional Development Fund. The Green Urban Living (GUL) is a multifunctional system based on expanded cork agglomerate for the construction of green roofs and living façades (**Figure 3**).

This system was developed within a national project, which aims to develop and validate new roof systems and green façades structured in expanded cork agglomerate (ICB), with a higher environmental and energy profile than conventional solutions and with a high capacity for energy customization and prefabrication. In these eco-designed systems, ICB will simultaneously provide: thermal insulation of the building; drainage functions; retention functions; and carbon capture.

Cork is a raw material that is so perfect that no industrial or technological processes have yet been able to replicate it. It has numerous advantages and benefits due to its key characteristics, and it represents an innovative opportunity to incorporate cork-based materials in green roof systems. Some of the cork's key characteristics are:

- Very light material—Over 50% of its volume is air, which makes it very light, it weighs just 0.16 grams per cubic centimeter and can float.

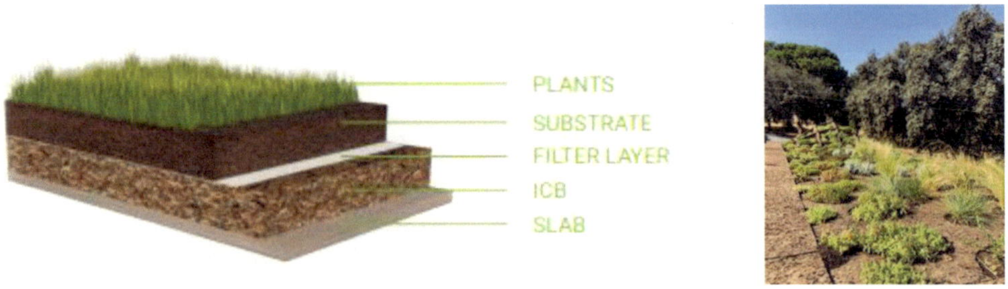

Figure 3.
Prototype layer scheme based on the GUL system (left). Example of an installation using the GUL system. Source: Neoturf (right).

- Elastic and compressible—It is the only solid that, when compressed on one side, does not increase in volume on another; and as a result of its elasticity, it is able to adapt, for example, to variations in temperature and pressure without suffering alterations.

- Impermeable to liquids and gases—Thanks to the suberin and ceroids contained in the cell walls, cork is practically impermeable to liquids and gases. Its resistance to moisture enables it to age without deteriorating.

- Thermal and acoustic insulator—Cork has low conductivity to heat, noise, and vibration. This is because the gaseous components contained in cork are enclosed in small impermeable compartments, isolated from each other by a moisture-resistant substance.

- Fire retardant—Cork is also a natural fire retardant: it burns without a flame and does not emit toxic gases during combustion.

- Highly abrasion-resistant—Cork is extremely resistant to abrasion and has a high friction coefficient. Thanks to its honeycomb structure, its resistance to impact or friction is greater than that of other hard surfaces.

- Hypoallergenic—Because cork does not absorb dust, it helps protect against allergies and does not pose a risk to asthma sufferers.

- Natural touch—The natural texture of cork combines softness and flexibility to the touch with a naturally uneven surface. The variable degree of irregularity is given by the type of cork used and the finish chosen.

The cork also plays an important role in Portuguese Economy, representing 2% of Portuguese total exports, and due to its sustainable, environmental, and economic benefits, it seems to be an excellent material to incorporate in the project, as well as the use of the GUL, in order to test it to the adaptation to a climate change in a school building.

3.4 Prototype for exterior: mBiGPond: exterior solution to promote water infiltration into the soil and reduce air temperature

It was important for LIFE-mybuildingisgreen project to address also the blue areas, from the perspective of water management, which may be due to the increase of the permeable areas (favorable to the infiltration of water into the soil), the reuse of rainwater, and the reduction of water for irrigation.

Following the solutions related to water, it would be important for the project to also contribute to the strengthening of this built urban relationship—permeabilized soil. In this context, urban ponds play an important role in the formation of small aquatic ecosystems of low demand and complexity.

The ponds are masses of standing water or very low current of a permanent or temporary nature, depending on the climate, the geology of the land, and the availability of water. The ponds are characterized by their low depth, total penetration of light into the water, possibility of occurrence of plants throughout their area, and absence of water stratification, and may originate from natural, geological or, ecological processes, or more commonly, as a result of human activities, intentional or not.

Ponds present themselves as a nature-based solution that act in water management in several dimensions: ensure the establishment of an aquatic ecosystem; ensure the availability of water (for fauna and flora); promote biodiversity; increase permeable area; boost water infiltration into the soil; and reduce the need for water for watering.

3.5 Facade prototypes in the pilot building in Badajoz. mBiFAVE 1 and mBiFAVE 2 systems

It has been decided to make systems superimposed on the façade that generate shadows in the gaps, preventing the direct incidence of solar rays, for this purpose canopies or cantilevered elements have been created, made up of NBS, plants, and plant species, in addition, so that the implantation of these elements does not prevent sunlight in the winter months, deciduous species have been used.

They consist of a removable modular substructure of laminated tubular frameworks, which support containers or pots with climbing plants, equipped with an irrigation system through pipelines adapted to the substructure. These systems are mainly constituted by a removable modular substructure of laminated tubular framework, of modules with dimensions 2x2 and 3 meters high, in hot galvanized steel profiles of 60.60 and 120.60, anchored to the façade and to the ground.

Formed by vertical elements (tubular steel pillars) and horizontal elements (crossbars and tubular steel beams), which support their own weight and that of the plant containers. It has been called FAVE (Vegetable Facade), in it, horizontal and vertical surfaces are generated on the holes and in front of them.

Two variants have been adapted in this system:

3.5.1 "mBiFAVE 1" systems

Containers or pots of guide plants, of the vine type, that upholster and generate surfaces that are opaque to the sun, will be arranged in the tubular framework. Two varieties of vines have been selected as guide plants: *Parthenocissus quinquefolia* (Virgin vine, originally from the USA) and *Parthenocissus tricuspidata* (Japanese vine or vine, originally from Asia). The system incorporates a mesh of galvanized steel wires that serves as a guide for plant growth (**Figure 4 left**).

3.5.2 "mBiFAVE 2" systems

Vegetated awnings are arranged in the tubular framework, continuous surfaces on racks on which species grow superficially on both sides (**Figure 4 right**).

The double vegetated awning is made up of:

- Rectangular frame of 0.80 × 2.90 mt, made with hot-dip galvanized tubular steel, anchored to the substructure.
- Leaf.skin tensioned membrane formed by anti-root PVC sheet fixed by linear clips on the frame.
- Hydroponic geotextile substrate.
- Lower gutter made of sheet steel, connected to a drain to collect excess water.

Figure 4.
Sketch of mBiFAVE 1" systems (left). Sketch of mBiFAVE 2" systems (right).

- Irrigation system is installed at the top from where the water falls by gravity soaking the entire substrate. The water provides the fertilizer, keeping the vegetation in perfect condition.

- Mixture of seeds that are projected on the geotextile, after a few months the membrane will be completely upholstered.

3.6 Cover prototypes in the pilot building in Badajoz. mBiCUVE 1, mBiCUVE 2, and mBiCUVE SUS systems

Three roof systems have been implemented, whose main function is to reduce the effects of solar radiation in the hot months and reduce energy losses in winter months and thus CO_2 emissions and energy consumption in heating, easy to implement and adaptation to any building.

3.6.1 System "mBiCUVE 1": (VEgetable Roof 1)

It is implanted on the existing flat roof, separated from it, generating an air chamber between both.

To do this, some racks are arranged, supported on concrete "plots", on which we place removable "trays" that house a thin extensive roof solution, with an improved substrate in which native species are planted. The system is made up of:

- Supports or plots: made in situ with concrete, with a height depending on the roof to achieve horizontal leveling of the frame.

- Removable modular frame: made by means of a framework of hot-dip galvanized tubular steel, with approximate dimensions of 4.5 m long x 1 m wide, formed by stringers and crossbars.

- Galvanized sheet steel trays: with side tabs for support (**Figure 5**).

3.6.2 System "mBiCUVE 2": (Vegetable Cover 2)

A system similar to the mBiCUVE 1 system, as a variant on the racks, large containers would be implanted, at ends and intermediate points, where the two varieties of vines that we have used for planting are planted. Façade systems, mBiFAVE 1.

3.6.3 System "mBiCUVE SUS": (Vegetable Cover with SUBSTRATE)

An extensive cover system is used with an improved substrate, including recycled aggregates for the realization of the cover drainage. A layer of protection, a drainage sheet, a filter, and the improved substrate in which native species are planted are placed on the waterproofing of the roof on which the action is being carried out.

4. Applications in Alentejo Central, Oporto and Badajoz

By analyzing the existing conditions, namely strong solar incidence in classrooms and the great thermal discomfort associated with high thermal oscillation, the intervention goes through the installation green wall that will work with vegetation curtain. Also, due to pathologies at the level of existing roofs such as poor waterproofing and nonfunctional pendants, green roofs of three typologies will be installed: extended sloped coverage, extensive coverage with gul system, and a green cover with solar panels—Solar bio roof.

With the concern of closing the water cycle, the surplus of rainwater from two sloping roofs will be routed to two tanks and later, by gravity, to a pond of natural infiltration, located in Horta das Oliveiras. Thus, the water will be returned to the natural soil, being able to recharge the aquifers.

For the monitoring and study of the ecosystem services of green roofs, equipment such as a weather station, various thermal sensors and humidity, and caudameters will be installed.

4.1 Green roofs prototypes

Green roofing has increasingly established itself as an important element in the strategies of "Nature-based Solutions" for Portuguese cities, thanks to the numerous environmental, financial, and social benefits they present (**Figure 6**).

Figure 5.
Sketch of mBiCUVE 1" system (left). The three roof systems: mBiCUVE 1, mBiCUVE 2, and mBiCUVE SUS (right).

For the success of the implementation of mass green roofs in cities and for their dissemination and acceptance by the professional groups involved and the general population, it is essential that there are examples of model projects promoted by the municipalities themselves as a demonstration of their advantages and promotion of good implementation and maintenance practices.

4.2 The mBi_GUL system

The GUL roof will be installed with the innovative green roofing system with expanded cork agglomerate—ICB. The cork agglomerate, with 8 cm thickness, will perform the functions of drainage of excess water, thermal insulation, protection of waterproofing, and sound insulation. Using cork as a building material—biomaterial—this system captures carbon in its production cycle.

The substrate profile will be 12 cm and herbaceous species will be planted in alveole. The planting strategy aimed to promote biodiversity in the green cover. Perimeter gravel strips 30 cm wide will be installed.

4.3 Solar bioroof prototype

The SOLAR BIOROOF green roof is a system that combines the installation of solar panels with the existence of vegetation. The substrate profile will have the thickness of 10 cm, where a pre-cultivated Sedum carpet will be installed. Perimeter strips of gravel with variable width between 22 and 30 cm will be performed.

The fixation of the solar panels will be carried out through solar bases with ballast, without drilling the slab, ensuring the tightness of the slab. The ballast will consist of 7 cm of river gravel. A counterweight life system is planned to be installed. This system consists of anchoring bases and steel cable, where the operators will have to secure themselves through sliding carts and harnesses. This system is also not intrusive, also ensuring the preservation of waterproofing.

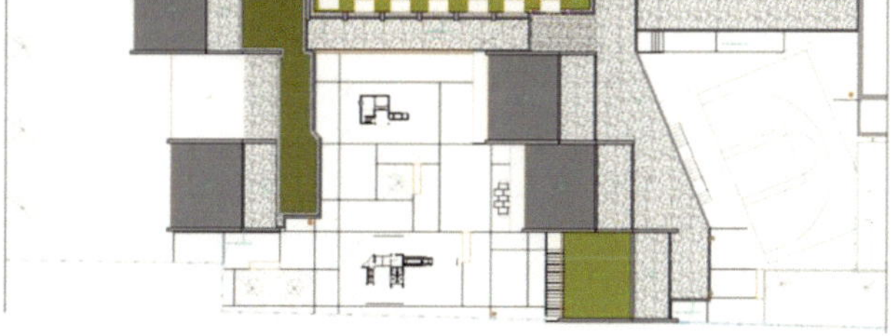

Figure 6.
Blueprint of green roofs in EB1 do Falcão. In different greens the 3 green roofs prototypes: GUL system, solar bio roof, and sloping green roof (ANCV—Associação Nacional de Coberturas Verdes).

4.4 Sedum carpet—sloping green roof prototype

The sloping green cover will be performed using the green roofing system up to 20° slope that holds supports to be installed along the lowest point of the cover. The substrate profile will be 10 cm and will be populated by the genus sedum in pre-cultivated carpet.

The excess water from the roofing system will be routed to a water tank.

4.5 Green façade

Green facades for a pleasant atmosphere. Putting vegetation on the facades is so far something that is usually left to chance. Attractive growth systems can be built for micro-vertical gardens with some easy-to-assemble components made of high-quality stainless steel.

Green facades are attractive, ecologically sensitive, and useful. They create a shadow area on the façade which makes the temperature more mild. In the case of the EB 1 do Falcão, due to the sun exposure and size of the windows, the thermal conditions felt inside the building cause discomfort during the warmer months, especially in the classrooms. To lower the temperature, it was suggested the implementation of a wall of steel cables covered by a deciduous vine.

The species selected is *Parthenocissus tricuspidata* because it is fast-growing, losing the leaves in winter, so that light can enter and heat the space so that it is more energy-efficient. This species changes its hue throughout the year, adding beauty and dynamism, transforming the landscape of the place with the seasons.

The vine is placed from 60 to 60 cm, being placed 10 feet along the façade, in a 50 cm wide construction site, supported by squares connected to the upper area of the façade from that extend stainless steel cables with a thickness of 6 mm and length of 6 meters. These cables have an intermediate attachment point on the balcony slab. As it is a small construction site, irrigation has been implemented using the drop-by-drop system.

The vegetation must be pruned to maintain its growth within the area assigned to it, at its base should be placed mulch and the weeds will have to be plucked.

The green wall structure should be composed of a cement or buried concrete base structure, about 20 cm. This 570 cm long piece will serve as a base anchorage for the steel cables installed by placing a 60-in-60 M10 eye. The 6 mm steel cable will be supported by two eyelets at each end, locking the cable with four 6 mm, two-wire cables at each end. Central fixation is placed through a third eye nailed to the building.

4.6 mBiG_Pond prototype

The construction of artificial lakes, creating water reserves for biodiverse purposes, are examples of fundamental hydraulic works in rational water management. In order to reduce the water losses of these structures, appropriate waterproofing techniques should be used (**Figure** 7).

The waterproofing of the pond is carried out with the placement of a geotextile fabric to prevent the rupture of the waterproofing screen. The water coming from two covers of the EB1 do Falcão, one with vegetation and the other without, is directed from the retention tanks to the lower level where the pond is located.

The vegetation proposed for this area will serve as a remediation plant, thus helping to maintain the quality of the water that is being infiltrated into the soil.

This space is of great importance because it creates a biodiverse area, where plants of riparian characteristics will arise and that small amphibians and insects can enjoy.

Both spaces will be areas of excellent learning opportunity for the students of the EB1 do Falcão, as they will be able to observe the natural dynamics present in this project, the change and cycles of nature, the emergence of new wildlife in the area of the pond, and the reduction of temperature inside their rooms, thus allowing greater thermal comfort.

5. Expected results and conclusions

With the execution of this project, the following results are expected:

- Climate adaptation of the pilot buildings (three educational centers) through the implementation of NBS prototypes.

- Well-being improvement and thermal comfort for around 1.000 citizens in the 3 pilot buildings and communities.

- The elaboration of Reference Reports and Good Practice Manuals on the application of NBS in public buildings of education services for the 3 climatic risk areas of the EU.

- For the entire Project it is expected to reduce the production of 27 Ton of CO2/ year and 144 kg of NOx (reduction 20% and 7% respectively).

- It will be collected about 2.700 m3/project of rainfall that represents 30% of each building.

- Increase in the area of green areas in each building by approximately 0.5 ha.

- Reduction of 50% of the energy costs for cooling and 10% for heating. This amount to 1000€/building.

- Integration of NBS into regulations, action plans, and environmental programs

- Execution of expert meetings, demonstration workshops, expert workshops, online seminars and transnational conferences, stakeholders training, website, publication of articles, connection with media associations, and videos.

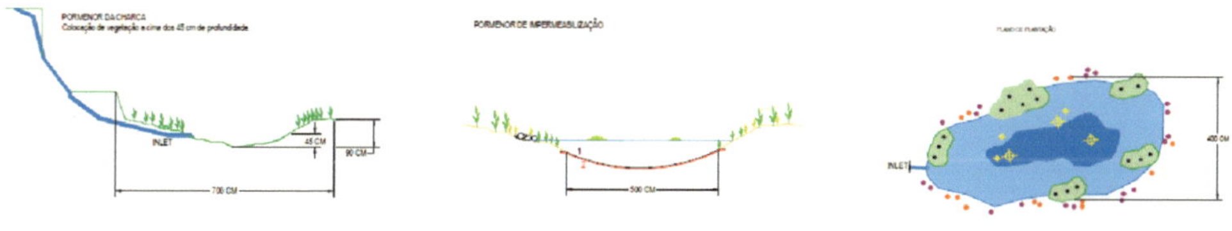

Figure 7.
Details of the mBiGPond—profile and top view of the pond.

The overall conclusion is that NBS can be a technical and scientific solution for this type of building allowing a more affordable and energy-efficient solution for the adaptation to climate change in Mediterranean regions.

It is already observed that the climate is changing, locally proved by the PIAAC study so it is time to act in order to use Mother Nature's knowledge to prevent more damage to Earth's environment.

Of particular relevance is the application in two study areas, Alentejo and Badajoz, since they correspond to two territories with the lowest population density and the lowest GDP in Portugal and Spain. In addition, these areas are suffering severe depopulation and economic crisis in their rural areas, so these technologies can help improve their social conditions.

Author details

Teresa Batista[1*], Salustiano Torre[2], José Fermoso[3], Nuno Morais[4], Ramona Avila[5], Carlos Pinto Gomes[6,7], Rui Alexandre Castanho[8,9,10] and José Cabezas Fernández[9,11]

1 CIMAC—Intermunicipal Community of Alentejo Central and MED—Mediterranean Institute for Agriculture, Environment and Development, Instituto de Investigação e Formação Avançada, Universidade de Évora, Évora, Portugal

2 Real Jardín Botánico, CSIC, Madrid, Spain

3 CARTIF Technology Center, Parque tecnológico de Boecillo, Valladolid, Spain

4 Câmara Municipal do Porto, Porto, Portugal

5 Diputación de Badajoz, Spain

6 Department of Landscape, Environment and Planning (DPAO), University of Évora, Évora, Portugal

7 Institute of Earth Sciences (ICT), University of Évora, Évora, Portugal

8 Faculty of Applied Sciences, WSB University, Dąbrowa Górnicza, Poland

9 VALORIZA—Research Centre for Endogenous Resource Valorization, Polytechnic Institute of Portalegre (IPP), Portalegre, Portugal

10 CNPQ Research Group Aquageo Ambiente Legal, University of Campinas (UNICAMP), Brazil

11 Research Group on Environment and Spatial Planning (MAOT), Universidad de Extremadura, Badajoz, Spain

*Address all correspondence to: mtfb@uevora.pt

References

[1] CIMAC. Plano Intermunicipal de Adaptação às Alterações Climáticas do Alentejo Central (PIAAC-AC) [Internet]. 2017. Available from: https://www.cimac.pt/sobre-piaac-ac/

[2] CEDRU / IGOT / WECONSULTANTS / CIMAC. Plano Intermunicipal de Adaptação às Alterações Climáticas do Alentejo Central (PIAAC-AC). [Internet]. 2017. Available from: https://cedru.com/o-que-fazemos/plano-intermunicipal-de-adaptacao-as-alteracoes-climaticas-do-alentejo-central/

[3] Schleussner CF, Menke I, Theokritoff E, van Maanen N, Lanson A. Climate Impacts in Portugal. [Internet]. 2019. Available from: https://youth4climatejustice.org/wp-content/uploads/2021/01/Climate-Analytics-Climate-Impacts-in-Portugal-min.pdf

[4] Cardoso RM, Soares PMM, Lima DCA. Mean and extreme temperatures in a warming climate: EURO CORDEX and WRF regional climate high-resolution projections for Portugal. Climate Dynamics. 2019;**52**:129-157

[5] Pérez FF, Boscolo R. Clima en España: pasado, presente y futuro. Informe de Evaluación del Cambio Climático Regional. Madrid, ESP: Ministerio de Ciencia e Innovación; 2010. p. 85

[6] Camara Municipal do Porto. Suporte Biofísico e Ambiente—Clima e Ambiente Urbano: Relatório de Caraterização e Diagnóstico. Plano Diretor Municipal. Porto, PT: Camara Municipal do Porto; 2018. p. 109

[7] CARTIF. Life—MyBuildingisGreen Technical report. [Internet]. 2019. Available from: https://www.cartif.es/life-mybuildingisgreen/

[8] Edenhofer O, editor. Climate Change 2014: Mitigation of Climate Change. Vol. 3. New York, USA: Cambridge University Press; 2015. p. 1435

[9] Akbari H, Cartalis C, Kolokotsa D, Muscio A, Pisello AL, Rossi F, et al. Local climate change and urban heat island mitigation techniques—The state of the art. Journal of Civil Engineering and Management. 2016;**22**(1):1-16

[10] Sotiris V, Chrysanthi D, Thornes J, Lai KM, Taylor J, Myers I, et al. Impact of climate change on the domestic indoor environment and associated health risks in the UK. Environment International. 2015;**85**:299-313

[11] Heracleous C, Michael A, Savvides A, Hayles C. Climate change resilience of school premises in Cyprus: An examination of retrofit approaches and their implications on thermal and energy performance. Journal of Building Engineering. 2021;**44**:103358

[12] Gómez G, Frutos B, Alonso C, Martín-Consuegra F, Oteiza I, de Frutos F, et al. Selection of nature-based solutions to improve comfort in schools during heat waves. International Journal of Energy Production and Management. 2021;**6**(2):157-116

[13] Gómez G, Frutos B, Alonso C, Martín-Consuegra F, Oteiza I, Castellote MM, et al. Prediction of thermal comfort and energy behaviour through nature-based solutions implementation. Case study in Badajoz (Spain). In: Hernández S, Chias P, editors. WIT Transactions on The Built

Environment. Vol. 195. Billerica MA, USA: WIT Press; 2020. pp. 17-27

[14] Fermoso J, Torres T, Antón MÁ, Peña A, Muñoz J, Torre S, et al. Improvement of classroom conditions and CO_2 concentrations through natural ventilation measures reinforced with NBS implementation. In: Ksibi M et al., editors. Euro-Mediterranean Conference for Environmental Integration. Cham, CH: Springer; 2021. pp. 2305-2309

[15] Ma N, Aviv D, Guo H, Braham WW. Measuring the right factors: A review of variables and models for thermal comfort and indoor air quality. Renewable and Sustainable Energy Reviews. 2021;**135**:110436

[16] Schnabel SC, Lavado Contador JF, Gómez Gutiérrez A, García Marín R. Aportaciones a la geografía física de Extremadura con especial referencia a las dehesas. Cáceres, SP: Asociación Profesional para la Ordenación del Territorio, el Ambiente y el Desarrollo Sostenible (FUNDICOTEX); 2010. 260

[17] European Commission. An EU strategy on adaptation to climate change—Communication from the Commission to the European Parliament, the Council, the European Economic and Social Committee and the Committee of the Regions. [Internet]. 2013. Available from: https://eur-lex.europa.eu/LexUriServ/LexUriServ.do?uri=COM:2013:0216:FIN:EN:PDF

[18] European Commission. Guidelines on developing adaptation strategies. [Internet]. 2013. Available from: https://eur-lex.europa.eu/legal-content/EN/TXT/PDF/?uri=CELEX:52013SC0134&from=EN

Chapter 8

Green Roofs Influence on Stormwater Quantity and Quality: A Review

Cristina Santos and Cristina M. Monteiro

Abstract

This chapter intends to make an extensive review of the influence that Green Roofs (GR) have on the quality and quantity of stormwater. These aspects are very important to define the benefits and the disadvantages of this nature-based solution that is being implemented worldwide to improve the sustainability of urban areas. Previous studies show that the characteristics of GR (such as dimensions, the composition of the different layers and the type of plants) have a major influence on the quality and quantity of the GR runoff. Despite the proven benefits in urban stormwater management, in some reported cases, the quality resulted worst and for some GR conditions, the effect on rainwater retention was minimal. They are key elements to make resilient cities so a clear understanding of their functioning and development is fundamental to avoid and minimize potential impacts of malfunctioning of these nature-based structures.

Keywords: green roofs, stormwater, rainwater harvesting, rainwater quality

1. Introduction

1.1 Water issues and nature-based solutions

Nowadays, people are using water more efficiently due to environmental awareness, tariff policies and technological innovations. In Europe, for example, a decreasing trend in water abstraction is being observed in the last decades and water consumption is decoupling from economic growth, as water use efficiency has increased in the related water-dependent sectors, such as public water supply and tourism. However, as stated by the European Environment Report nr. 12/2021 [1] the issue of water stress continues to escalate as climate change exacerbates seasonal variations in water availability. The observed decreasing trend in water abstraction volumes has so far not translated into an improvement in the quantitative status of water bodies, and this may be partly due to the slow process of recovery and also to climate change, which can offset volumetric gains and aggravate local pressures. On the other hand, the urban population continues to grow, which leads to an increase in urban water demand. To continuously supply this demand, new storage systems, abstraction and networks are developed but with serious consequences to freshwater sources and ecosystems. Urbanization brings also soil sealing and flood situations, once the amount of naturally infiltrated stormwater is significantly reduced. The variability and pressure caused by climate change and seasonal variation in water supply tend to aggravate these problems. Thus, approaches focusing on valorizing unconventional resources (e.g., desalination, water reuse, rainwater harvesting) are already implemented in many Member states. Besides, several European Union policy initiatives support the use of Nature-based Solutions (NbS) to reduce Europe's vulnerability to water stress [1].

NbS are systems inspired and supported by Nature, implemented in urban areas to restore vegetation and natural ecosystems. The Federal Emergency Management Agency (FEMA) of the USA describes them as sustainable planning, design, environmental management, and engineering practices that weave natural features or processes into the built environment to build more resilient communities [2]. On a neighborhood or site scale, they are defined as distributed stormwater management practices that manage rainwater where it falls. These practices can often be built into a site, a corridor, or a block without requiring additional space.

The use of NbS infrastructure can reduce the cost of stormwater management for new development because material costs and land disturbance are lower than traditional drainage [2]. The use of Nbs can also help to control stormwater that enters combined sewer systems, resulting in important savings in the amount of water treated in wastewater treatment plants and the consequence of flood occurrences. As referred by FEMA [2], New York City developed a plan to reduce combined sewer overflows using green and gray infrastructure and they predict that the NbS component will capture runoff from 10% of the impervious areas with a cost of about $1.5 billion, while the gray infrastructure option would cost about $3.9 billion.

The European Commission [3] states that implementing NbS on a larger scale would increase climate resilience while contributing to multiple Green Deal objectives. Buildings can contribute to large-scale adaptation, for example through local water retention and urban heat island effect reduction when incorporating green infrastructure.

NbS include green infrastructure technologies such as Green Roofs (GR), which are being implemented as part of a combined sewer overflow abatement strategy and to develop co-benefits of diminished stormwater runoff, including decreased loading of contaminants to the wastewater system and surface waters [4]. The study made by Koc et al. [5] revealed the important contribution of GR to stormwater management in urban areas: among different solutions (bioretention cells, permeable pavement and infiltration trench, isolated or in combination), GR provided the highest improvement with approximately 40% in both peak discharge and volume reduction and has been found as the optimal practice among the stand-alone solutions. The main reason for those results is that GR provides the closest characteristic to a natural basin by capturing the rainfall and increasing the time that a water drop takes to reach the drainage system. The best-integrated evaluation showed that GR and bioretention cells can significantly reduce the impacts of urban flooding.

Transitions towards urban built environments that incorporate sustainable drainage systems encounter several barriers such as limited evidence on their performance or concerns about negative impacts that they might cause. Collective efforts from different stakeholders, are essential to highlight the important benefits provided by them [6].

1.2 Aims and scope

This chapter intends to present the effects of GR in stormwater management. This NbS is being implemented worldwide and its effects on stormwater quality and quantity are being intensively investigated. These structures have an important impact on stormwater control and reduction, but they change the quality of the drained water. On the other hand, their efficiency is highly influenced by their characteristics and therefore, the more relevant design aspects should be carefully assessed. Intense literature research was made to analyze these questions. GR are characterized in detail in sub-chapter 2, and their effects on stormwater quantity and quality are detailed in sub-chapter 3, based on the main conclusions of experimental activities developed around the world.

The bibliographic research was made in two steps. In the first one, the "science direct" database was used, with four keywords: "green roof" stormwater quantity; "green roof" stormwater quality; "green roof" rainwater quantity; "green roof" rainwater quality. The advanced search (including title, abstract or author-specified keywords) was used and a total of 91 articles were found, 24 of them were duplicated. In the second step, the abstracts of the remaining 67 articles were analyzed and 24 of them were categorized as out of scope. Besides, 3 articles were not considered because two were extended abstracts from conferences and one was not available. The final result of the bibliographic research was 40 articles, published from 2007 to 2021. They were fully read to sustain the values and ideas presented in this chapter.

2. Green roofs

GR (also known as vegetated roofs) are defined as any type of soil–vegetation system established on building floors or roofs excluding the cases of pot vegetation [7]. In constructive terms, they are implemented in the building structure with the following components, from bottom to top: a high-quality waterproof membrane, a root barrier to protect the membrane, a drainage layer, a growing substrate and finally, the plants (**Figure 1**).

The GR vegetation most commonly used worldwide is Sedum, which has attributes of succulent plants such as a low growth rate and drought resistance that leads to low maintenance requirements [9]. Besides, using different plant species and

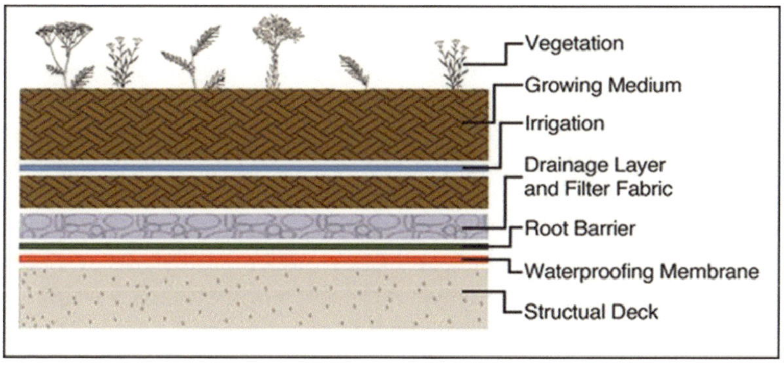

Figure 1.
GR section (the irrigation system is optional) [8].

creating diverse ecosystems, is generally more successful than planting single varieties of vegetation. Liu et al. [10] presented vegetation layers with three different plant species - Radix Ophiopogonis, Sedum Spectabile and Iris – with high resistance abilities to drought stress conditions and widely incorporated in local landscaping works and GR designs. Kuoppamäki [11] studied GR with two levels of establishment method: either by planting with seeds and plug plants or as pre-grown mats. The total thickness of the growing substrate of all meadow roofs was set to 10 cm. Meadow roofs established on-site with plantings are dynamic ecosystems, with wide temperature variation, which contribute to high evapotranspiration, resulting in lower moisture at the start of rain events. Although, this difference declined with time, as roof ecosystems evolved, the author considered plantings a better design than mats considering various factors. Guidelines were already developed by many countries to recommend the appropriate selection of plants to improve the benefits and to overcome difficult conditions prevailing on the rooftop, such as water scarcity and extreme climatic conditions [12].

Growing medium (also referred to as the substrate layer) is an essential component of GR that directly influences plant growth and also benefits GR regarding stormwater management, rainwater buffering, building energy savings, and sound insulation [12]. Substrates for GR are typically a mix of inorganic sand, organic soil, and fertilizers, which differ depending on the vegetation used. Sand, white charcoal debris mixed with sand, organic matter mixed with soil, and burned reservoir sludge mixed with rice hulls are examples of substrate constituents, that can also include the use of recycled materials [9]. Different ratios of organic and inorganic constituents result in GR substrate with distinct characteristics, appropriate for different applications.

The drainage layer maintains a non-water-logged and aerated condition for the substrate to support healthy plant growth. It plays a vital role in a GR, also by storing water between the substrate and the drainage material (or in the compartments of drainage plastic modules) that is used by plants in dry periods [12].

The root-barrier is optional for extensive GR but highly recommended in intensive ones once it protects the waterproof membrane from the roots of large plants. When necessary, an insulation layer might also be considered, as it prevents the water retained in the GR from extracting heat in the winter or cooling the air in the summer.

The waterproofing membrane will prevent the moisture of the substrate and drainage layer to pass into the structural deck, so its high quality and well execution are fundamental to keep the roof in good conditions and to avoid humidity problems inside the buildings.

GR systems may be modular, with drainage layers, filter cloth, growing medium, and plants already prepared in movable, often interlocking grids, or loose laid/built-up whereby each component of the system may be installed separately [8]. As referred, they present distinctive characteristics in terms of growing medium and plants but are usually categorized due to their intensive or extensive planting as stated by Berndtsson [7]:

- Intensive GR have deep soil layers and can support large plants and bushes. Typically, they require maintenance in the form of watering, fertilizing, and weeding;

- Extensive GR have thinner soil layers and smaller plants which in the final stage are expected to provide full coverage of the roof. These GR are more commonly used because they can be implemented in existing buildings and are generally maintenance-free. Some fertilization is often recommended especially in the first years.

It should be noted that distinct authors differ in the limits of substrate layer thickness defined for intensive and extensive systems, but the majority defines the 0.15 m as the frontier. **Table 1** resumes the main characteristics of both types of GR.

Existing full-scale GR, reported in the literature, present the following general characteristics:

- In Helsinki and Espoo, Finland: nine thin-layered (2–12 cm of the substrate) Sedum/moss roofs, and two meadow roofs with substrate depths of 21 and 23 cm [13];

- In New York, USA: a GR composed by six types of native sedum located over 10.2 cm of lightweight growth media. Above the growth, media is a drainage composite, with a water retention capacity of 7.9 mm of water, underlain by an adhered, waterproof, single-ply, thermoplastic polyolefin membrane. The membrane was installed over moisture-resistant gypsum board, tapered roof insulation, vapor barrier, and concrete deck [14];

- In Manchester, UK: a 43-year-old intensive GR with an average depth of 170 mm. It has standard construction with the vegetation and substrate layers divided from the 'egg box' design plastic drainage layer by a fibrous membrane. The roof itself is protected by a tough geotextile membrane. It has a mineral soil substrate rather than the more usual, prefabricated, lightweight aggregate (LWA) based substrate [15];

- In Alberta, Canada: an extensive GR planted with nine different vegetation species including three sedum varieties, two types of grass and four different flowering forbs. The GR growing medium are composed of recycled materials and minerals, and enhanced with compost, presenting a depth of 150 mm. On the bottom, the GR includes a Floradrain FD 25-E drainage board (with holes facing up) with a separate filter sheet [16].

In a conventional GR, such as the one represented in **Figure 1**, part of the rainfall is stored in the substrate layer and consumed by plants, another part returns to the atmosphere by evapotranspiration and the remaining rainfall is discharged as runoff. The amount of rainfall that is not released as runoff is affected interacts with the components of the GR and their physical environment and is defined as retention [11]. The dual-substrate-layer GR presented by Wang et al. [17], consisting of an upper organic nutrition layer for plant growth and a lower inorganic adsorption layer for water retention and pollutant reduction. This new generation of multilayer GR also present an additional layer to store rainwater which, if properly treated, can be reused for different purposes [18]. Cost analysis proved the practicability of dual-substrate-layer GR in retaining rainwater, but their long-term rainwater runoff quantity and quality performance in urban environments merit further investigation [17]. Xu et al. [19] studied a hydroponic GR system, developed to reduce urban stormwater runoff and to collect, treat and reuse greywater and rainwater onsite in green buildings. According to the authors, this system has a greater potential to reduce urban flooding relative to traditional GR and ensures the long-term effectiveness and stability of water reuse by intercepting rainwater and achieving long-term collection and treatment of greywater.

	Intensive GR	Extensive GR
Growing medium (substrate layer)	>15 cm	<15 cm
Weight	Heavy; require concrete support	Low; used with concrete, steel and/or timber support
Plants	Perennials, lawn, shrubs, small trees and rooftop farming	Grass, herbs, mosses, sedums and other succulents
Irrigation needs	Constant irrigation using an automated sprayer	Required only in dry periods
Maintenance	High	Low
Fertilization	Required trough the GR lifetime	Not required (except in the first years)
Waterproofing	High-end	Low-end
Root-barrier	Required	Not required
Overall costs	Higher	Lower

Table 1.
Characteristics of different GR (adapted from Vijayaraghavan et al. [12]).

The most relevant benefits associated with GR, are: (1) regulation of the urban thermal climate (minimizing urban heat island effect) and reduction of the energy consumption in buildings, (2) retention, reduction and quality improvement of stormwater, (3) enhancement of air quality, (4) creation of habitat for flora and fauna, (5) and supporting elements related to physical and mental well-being and (6) esthetic improvement of the urban environment [20, 21]. GR can also contribute to the food supply in urban areas, specifically the intensive ones, if they were used as rooftop farming. Apart from the great influence on the esthetics of the buildings, the study reported by Karteris et al. [22] showed that spices and aromatic plants achieve high CO_2 sequestration rates in extensive GR. They also showed their potential of reducing CO_2 emissions due to energy conservation and CO_2 absorption by GR. The energy conservation rates are strongly connected to the date of construction of the buildings, the available roof area and the number of floors and the use of the building. Large-scale implementation of GR can improve the mean rainwater retention rate in cities and thus contribute to the reduction of rainwater runoff and flood occurrence.

The survey made by Chen [9] in Taiwan revealed that the major concerns in selecting the type of GR, were the drainage performance and potential damages on the roof structure. People worried about litter clogging the drainage pipes and causing overflows on the roofs, and about water leakage on the ceilings due to the intrusion of the plant's roots into the roof construction materials. Thus, the most existing GR in Taiwan were the extensive type. The spatial distribution of GR benefits is dependent on the spatial distribution of suitable roof areas and environmental stressors. For example, the case study of the inner-city area of Braunschweig (Germany), presented by Grunwald, et al. [23], had great potential, with a higher percentage of benefits in comparison to residential areas. This is very relevant for this type of green infrastructure, that can be implemented on already existing roof areas (usually called as retrofit). GR is a highly sustainable and efficient solution to introduce additional vegetated areas within densely built cities where the impact of environmental stressors is usually high.

Despite the benefits of GR in managing urban stormwater quantity and quality, several studies have demonstrated that GR can pose negative impacts on the urban environment due to chemical leaching, in particular in their early age [16]. They are a living system with many components, so understanding its dynamics is necessary to predict potential environmental impacts. Also, a significant challenge in their widespread implementation is the execution costs, which are higher when compared to traditional roofs. However, the difference can be attenuated if the benefits are well investigated and documented. As referred above, besides the improvement in stormwater control and thermodynamic performance, the vegetative cover of GR also brings an extra source of CO_2 capture into the cities, and provides habitats for wildlife.

3. Impacts of green roofs on stormwater

As urban areas continue to expand, stormwater negative consequences are increasing, and climate change will worsen this situation. In this scope, particular focus should be given to cities with combined sewer systems where stormwater and wastewater are conveyed to the same pipes. In such cases, additional costs are allocated to the treatment plants and, when heavy rain events occur, the affluent volume can exceed the capacity of the system and raw untreated sewage might flow out of relief points into water resources.

A GR changes stormwater runoff, when compared with that from a traditional hard roof, through lowering and delaying the peak runoff. For the same rain event, the peak from the traditional hard roof occurs sooner because a certain water volume is detained in the GR multilayer system and a part of it is retained, as explained previously. The retained water is evaporated and transpired by the plants, and this explains the observed runoff volume reduction [7]. Besides, while flowing through the GR, the water interacts wilt the different organic and inorganic elements and changes its quality. Thus, GR have a direct influence on the stormwater runoff, mainly due to the characteristics of their components. The magnitude of this influence and the factors that contribute the most, both to the quantity and to the quality of the runoff, are described in the following subchapters.

3.1 Effects on stormwater quantity

For retaining water, the GR substrate texture and porosity act as a series of pipelines that store water and control the flow rate at which water runs through the growing medium. Before reaching its peak capacity, the runoff from a GR is lower than the runoff from a traditional roof. While water flows through the substrate, it is partially consumed by plants or retained in the substrate porous, so the total runoff volume at the end of rainfall events is also lower for GR than for traditional roofs. The retention capacity of a GR is the difference between the runoff volume of a conventional and GR and that difference is bigger in small-intensity rain events, considering that the substrate is not saturated. After reaching the peak capacity or if the antecedent dry weather days (ADWD) are not enough to restore the retaining capacity of the substrate, the benefits of the GR concerning stormwater control are minimized.

The reduction in GR runoff generally ranges from 50–100% depending on the distinct factors described below [24]. Once the runoff is released over a longer period of time, these structures decrease the flooding risk and improve the performance of the networks downstream.

Kuoppamaki [11] quantified the capacity of Meadow roofs to retain and detain runoff in a field experiment in southern Finland, with varying weather conditions over 4 years. The meadow roofs showed a 40–70% annual cumulative retention of rainfall. Three main factors influenced this retention: (1) the seasonal variability, as the highest retention was measured in the rainiest season, (2) the evapotranspiration, because the retention capacity declined for their study with the development of plant cover and (3) the establishment of root systems that may have created preferential flow paths that reduced storage capacity.

Vegetation plays an important role in the retention and detention capacity of a GR. Despite the 100% vegetation cover of the mat's treatment studied by Kuoppamaki [11], these roofs retained less rainfall annually (40–60%) than less densely vegetated plantings (50–70%). According to Liu et al. [25], trees and shrubs, when compared to grasses, have a higher capacity to retain stormwater, due to: (1) creation of a denser canopy and trunk layers, which increases the evaporation area and trunk flow; (2) establishment of a thicker litter layer on the substrate surface, with excellent water-holding capacity, that extends infiltration time and reduces surface runoff; and (3) more abundant roots with a higher water-retention capacity. Also, the nine-month GR pilot study performed by Harper et al. [26] showed an approx. 40% reduction in runoff from the unplanted growing medium and an approx. 60% reduction in runoff from the planted growing medium. They highlighted the seasonal impact of plant evapotranspiration as the reduction in runoff was dependent on the season. Plants were a statistically significant factor in fall and spring, but not in winter once in wet seasons a reduced overall impact of growing medium and planting on total flow was registered. Authors stated that plants had a 20% additional reduction in stormwater runoff in the fall, despite relatively recent establishment, and when the plants were dormant over winter, less variation between the planted and unplanted trays was observed, as would be expected under low evapotranspiration conditions.

Due to the dynamic of water flowing through the GR, the depth of the substrate and its initial moisture have a major influence on precipitation retention. Lee et al. [27] showed that a GR with a soil depth of 200 mm (intensive GR) reduced runoff by 42.8–60.8% compared to a 13.8–34.4% reduction of a GR with a soil depth of 150 mm. Viola et al. [28] also explored the retention performance of GR as a function of their depth and in different climate regimes, using both an intensive and an extensive GR. The amount of retained water increased in higher substrate depth, because more water was allowed to be stored in the active layer and consequently evaporate from the system. These results lead to the need to focus on real, intensive GR to characterize their benefits. In that scope, the aged intensive GR studied by Speak et al. [15] in Manchester, UK, achieved average retention of 65.7%. A comprehensive soil classification revealed that the growing substrate, a mineral soil used, was in good general conditions and also high in organic matter content which can increase the water-holding capacity of the soil.

Another factor that can influence the water retention and detention performance of GR is the irrigation. Perales-Momparler et al. [6] highlighted the importance of planting with vegetation with low water needs: during the start-up period of the green roof, irrigation was carried out to ensure the proper establishment and development of the plants, which significantly reduced its hydraulic efficiency to volumetric efficiencies of up to 50%. When irrigation operations were less frequent (in winter and spring seasons), the volumetric efficiency raise. In general, their study showed significantly different hydraulic behavior between GR and a conventional roof:

during a typical short torrential shower (total rainfall volume of 29 mm and the maximum 10-min intensity of 43 mm/h) showed only 26% of the rainfall volume detained by the conventional roof whereas 86% efficiency was achieved in the green roof. The peak flow reduction was also substantial: conventional runoff was seven times greater than that from the green roof.

The hydrological performance of the dual-substrate-layer referred above (with a lower inorganic adsorption layer for water retention and pollutant reduction, studied by Wang et al. [17]) was relatively good because of the porous structure and large specific surface area of the adsorption materials employed, which could help to absorb and hold the water in the GR structure. The rainfall retention values observed for the six dual-substrate layer GR in their study varied from 21.1% to 81.9% with an average of 48%. The different compositions of them explain the distinct retention values. The dual-substrate-layer GR, which used the mixture of activated charcoal with perlite and vermiculite as the adsorption substrate, possessed better rainfall retention performance.

Due to the limited retain capacity of GR, the size of the rain events also influences significantly the hydrological performance of all types of GR. Carpenter et al. [4] made a study based on rainfall events to evaluate annual retention during the growing and nongrowing seasons of an existing extensive GR, composed of six types of native sedum. No differences in water retention of the GR for the growing and nongrowing seasons were found but it was inversely dependent on the size of the rainfall. Under low rainfall inputs, the GR was partially saturated and retained high amounts of rainwater (between 98% and 100% of the incoming precipitation). Inversely, high-intensity rain events and a decrease in the antecedent dry period, promoted saturation of the roof, which decreased its retention capability to only 88%. The 150 mm of soil layer GR studied by Zhang et al. [29] (with retention capacities ranging from 35.5–100%) showed a significant negative relationship between the depths of rainfall and the runoff retention rates. However, there was no correlation between the runoff retention and the antecedent dry period. The size (or depth) of the rainfall is a major factor also reported by other authors, that presented lower retention for high rainfall events [11] and a high retention efficiency for small magnitude rainfall events [26, 27].

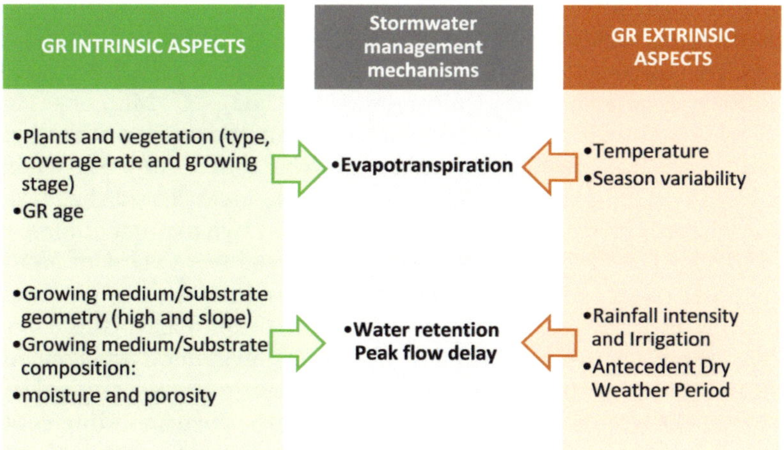

Figure 2.
Hydraulic performance of GR: intrinsic and extrinsic influencing factors.

Viola et al. [28] add that the performance of a GR increases when rainfall and potential evapotranspiration exhibit the same seasonality during the hydrological year happens, such as in humid subtropical climates. Conversely, the GR presents the minimum efficiency when rainfall and potential evapotranspiration are in counter-phase, as is found in a Mediterranean climate.

Figure 2 resumes the main aspects that influence the hydraulic performance of GR, from internal and external point of view.

3.2 Effects on stormwater quality

GR also change the qualitative characteristics of stormwater runoff, due to the interaction between water and the different structural components. Both plants and substrate are expected to have a direct influence on the runoff quality, but it can be positive or negative, as they can filter or eventually be a source of contaminants. Besides, considering that rainwater is relatively pure (apart from being acidic and containing traces of nutrients and metals) the magnitude of the pollution caused by the GR can be very high [12]. The quality of the drained water flowing from GR has been carefully investigated in the last decade which is extremely important as cities are becoming greener. However, if the impacts on stormwater quality are not well known while GR are being promoted and spread over cities, there is a risk of providing contaminated water to the receiving watercourses.

It must be taken into account that GR vegetation may trap airborne particles and dust, thus eliminating them from the rainwater, which is beneficial. Also, the substrate material can act as an ion-exchange filter for pollutants, nutrients, and trace metals present in rainwater. If the ion concentration of the rainwater is high the GR components may act as a sink, decreasing the ion concentration in the GR runoff. Otherwise, if the concentration is significantly lower than the substrate one, some ions will seep into the rainwater from the substrate resulting in a higher ion concentration in the drained rainwater [12, 30].

Gregoire and Clausen [31] studied the runoff quantity and quality of a 248 m^2 extensive GR and a control roof in Connecticut, USA, using a paired watershed study. The mean concentrations of TP (total phosphorus) and PO_4–P (orthophosphate as phosphorus) in GR runoff were higher than in rainwater, but lower than in the runoff from the control roof. They concluded that the GR was a sink for NH_3–N (ammonia-nitrogen), Zn (zinc), and Pb (lead), but not for TP, PO_4–P, and total Cu (copper). The GR also reduced a load of Total Nitrogen (TN), Total Kjeldahl Nitrogen (TKN), nitrate and nitrite nitrogen (NO_3 + NO_2 –N), Mercury (Hg), and dissolved Cu due mainly to the reduction of the stormwater volume. The growing medium and slow-release fertilizer were the probable sources of P and Cu in GR runoff, pointed by the authors. Berndtsson et al. [32] also investigated the influence on runoff water quality from two full-scale GR located at two distinct climate regions (Japan and Sweden). Both had similar performance and were a sink of NO_3–N and NH_3–N. They also release dissolved organic carbon and potassium. The intensive GR was a sink of TN and Phosphorus was released from the extensive one. They were generally not a significant source of metal pollutants. In general, the concentrations of nitrogen and phosphorus compounds, as well as heavy metals, studied in the runoff from the GR were in a similar range or below the corresponding concentrations in the urban runoff. The authors concluded that GR did not provide the benefit of rainwater treatment (with exception to nitrate nitrogen) but neither would other urban impermeable surfaces.

The most common contaminants of urban stormwater runoff are heavy metals, petroleum hydrocarbons, pesticides, suspended solids, nutrients, and pathogenic microorganisms. Pollutants studied in GR runoff are most often some forms of phosphorus and nitrogen, and heavy metals. The factors that most influence GR runoff quality can be listed as follows [7] and described in the following paragraphs:

- type of material used (composition of the growing substrate, drainage material and/or underlying hard roof material, rain pipes material);
- type of vegetation, season (biomass using nutrients);
- soil high;
- type of drainage;
- maintenance/fertilizers used;
- dynamics of precipitation;
- wind direction;
- local pollution sources;
- physico-chemical properties of pollutants.

Liu et al. [30] studied GR with different substrate material compositions, substrate depths, vegetation types and slope gradients, and concluded that they acted as a source of most of the studied water quality parameters. Substrate materials and vegetation types contributed notably to water quality and the ones with the less organic materials in the substrate showed better water quality. The influences of substrate depths and slope gradients on runoff water quality were small. They noted that the mean electrical conductivity (EC) increased from 295.17 μs/cm in rainwater to 418.33 μs/cm in GR runoff, due to the increase in dissolved ion content. The substrate composition was the pointed cause of the elevated concentration of anions and cations in the runoff. Still, the majority of water quality parameters of GR outflows were partly considered to be good. Previously, the same authors had studied the influence of substrate and vegetation on the water quality of GR outflows, by designing a scale-based runoff plot of extensive GR with different substrate and vegetation types [10]. The results showed that the TSS (total suspended solids), TN (total nitrogen), and TP average concentrations of the GR runoff were all significantly higher than that of the conventional roof runoff. The TN and TP concentrations of 5 cm substrate depth were significantly lower than that in the 15 cm substrate depth. This result could be explained due not to the depth of substrate but the higher content of organic matter in its substrate, thus the nutrition was more prone to being leached out from the system. In this specific case, significant differences were observed in TN concentrations between the vegetation species Sedum Spectabile and Radix Ophiopogonis, but no significant differences were observed about TP concentrations. These results imply that the phosphorus in the substrate might not be used efficiently by plants and can be leached out by stormwater runoff. Beecham and Razzaghmanesh [33] also highlighted the importance of vegetation in enhancing pollutant removal in GR systems. In the study comprising sixteen, low-maintenance and unfertilized intensive and extensive GR testing beds (100 mm and 300 mm depth media, respectively), the intensive GR performed better than the extensive ones about outflow water quality, while in the non-vegetated beds, the extensive beds performed better than intensive systems. In addition, growing medium with less organic matter had better water quality performance. Generally, the pollutant concentrations were higher in runoff from non-vegetated beds. This was believed to be partly due to leaching from the growing medium and partly due to plant uptake in the vegetated beds.

However, the results reported by Akther et al. [34] pointed in the opposite direction, because vegetation played a secondary or minor role when compared to growing substrate in the nutrient leaching of the GR outflow. They characterized and modeled the temporal behavior of nutrient leaching from GR in both laboratory and field settings and concluded that the leaching behavior was very similar in both locations: the degree of nutrient leaching declined temporally, and the growing medium was proved to be the primary source for nutrients being gradually washed off by infiltrated water. In addition, they showed that nitrogen (N) leach quicker than phosphorus (P) from the GR, which is consistent with the higher mobility in the soil of N compared to P. The degree of nutrient leaching in their study was associated with the chemical properties (e.g., the nutrient content) of the growing medium.

Razzaghmanesh et al. [35] also studied the growing medium influence in the water quality of the outflow from intensive and extensive full-scale GR located on the roof top of a 22-storey building in the Adelaide CBD, South Australia. The values of the parameters such as pH, turbidity, nitrate, phosphate and potassium in intensive GR outflows were higher than in the outflows from the extensive GR, (except for some events and EC, TDS (total dissolved solids) and chloride). Generally, the performance of the extensive GR was better than the intensive systems in terms of pollutant removal, which may be related to the reduced volume of substrate that can leach pollutants. A very interesting aspect highlighted by the authors was that the contaminant concentrations in runoff from both intensive and extensive GR generally decreased during the study period: the concentrations at the start of the investigation were higher than towards the end of the nine months for most of the studied pollutants. This was an expected result once the systems settle down and the chemicals within the growing medium leach out over time. **Table 2** presents a resume of GR and traditional roofs runoff values.

To understand the influence of roof age and hydro-meteorological variables on the temporal evolution of chemical leaching in GR, Akther et al. [16] monitored in 2015–2018 a full-scale extensive GR and a reference roof located in a cold and semi-arid climate region. During the study period, their GR also leached nutrients (N and P) and constituents represented by EC, whereas it removed metals (Zn, Cu, and Pb). The degree of chemical leaching declined temporally: the leaching of N appeared to cease, whereas P leaching was still ongoing at the end of the study period. Also focusing on the GR age, Todorov et al. [36] presented a four-year study of water quality in runoff from an extensive vegetated roof, sedum covered, initiating with the first growing season. The GR also proved to be a sink of N, TP and chloride (Cl), and a source of phosphate and dissolved inorganic and organic carbon. Once the roof growing medium is designed to sustain plant growth, it is supposed to be rich in N and P. Both NO_3^- and NH_4^+ are essential nutrients for the plants and their retention in the vegetated roof was possibly a result of assimilation by the sedum species. In general, the water drained

Water quality parameter	Experimental data		
	Asphalt roof (average, mg/L)	Aluminum roof (average, mg/L)	Green roofs range (mg/L)
pH	7.13	7.58	6.72-8.45 [b]
Turbidity (NTU)[a]	2.98	1.26	4.0-300 [b]
EC (μS/cm)	75.00	23.00	<100 [b]
Chloride (mg/L)	1.25	1.03	<30 [b]
Total dissolved solids (mg/L)	37.00	31.00	385.77
Nitrate (mg/L)[a]	2.62	1.90	2.20-39.20
Phosphorous (mg/L)[a]	0.14	0.16	0.20-2.20
Potassium (mg/L)[a]	9.60	3.00	38.37

[a]*Pollutant level is above the potable standard level.*
[b]*USEPA. Guidelines for water reuse (2012) EPA/600/R-12/618.*

Table 2.
Water quality of outflows from green roofs and control roofs (adapted from Razzaghmanesh, et al. [35]).

from the vegetated and impermeable roofs presented good quality, meeting the United States Environmental Protection Agency freshwater standards for all parameters, except for TP.

The studies presented so far, might suggest that GR are bad solutions regarding stormwater quality, at first sight. The work presented by Perales-Mompaler et al. [6] is no exception, once the resulting values of chemical oxygen demand (COD), TSS, TN and TP were higher than the non-vegetated roof, except for TSS which was usually below 20 mg/L. Nutrients (TN and TP) and organic matter (COD) were notably higher in runoff from the GR showing the washing of dissolved substances. However, despite presenting higher concentrations of COD and nutrients, the total loads drained by the GR were lower than that drained by the non-vegetated roof, due to the higher volumetric efficiency of the green roof. Also in this study, the reported washing effect declined after some time: COD concentrations decreased from values higher than 350 mg/L to 50 mg/L and similar trends were observed for TN and TP. After 17 rainfall events (total volume drained 9.0 m^3), TN and TP concentrations were reduced by approximately one-half. Carpenter et al. [4] examined a GR in Syracuse, NY and also reported that overall, nutrient losses were low because of the strong retention of water. They reported seasonal variabilities, as runoff waters exhibited a high concentration of nutrients during the warm temperature growing season, particularly TN and dissolved organic carbon (DOC). There was marked variation in the retention of nutrients by season shown by the variation in roof runoff concentrations. The GR served as a sink for wet deposition inputs of TN and TP for the majority of rain events. In contrast, the roof was a source of DOC during the growing season and a sink of DOC during the non-growing season. Overall, the experiment also revealed that nutrient retention is dependent on water retention. Although, nutrient concentrations and fluxes in roof drainage were greater than in the incoming precipitation, the total loading of TN and TP was attenuated because precipitation quantity was strongly retained. According to the authors, the elevated concentrations of those nutrients in GR drainage are not an issue of concern as the strong retention of water results in limited mass loss of nutrients from the GR to the sewer system.

A direct correlation exists between the magnitude of the rain events and the number of solids in the GR runoff. During small rain events, nutrients and sediment that are washed off from traditional roofs are retained on the GR because there was no runoff. This would lead to misleading higher concentrations from the GR in future larger rain events. New GR tend to be a source of pollutants, due to the initial nutrient load that comes from the decomposition of organic matter that was incorporated into the original substrate mix. On the opposite perspective, established vegetation and substrates can improve the water quality of runoff by absorbing and filtering pollutants [24]. These findings lead to more diligent use of fertilizers to minimize the adverse impact on stormwater runoff. Even though, if nutrient loading remains a problem on implemented GR a possible solution is to couple them with other green infrastructures, such as rain gardens and bioswales.

Buffam et al. [37] analyzed the runoff from more than 80 rain events over 2 years, in the extensive GR in Cincinnati, USA referred previously. Strong seasonal patterns were observed in bioactive elements. Carbon, nitrogen, phosphorus, and base cation concentrations were highest in the summer, and were positively correlated with temperature. They concluded that temperature-mediated processes, rather than plant uptake or hydrologic variation among storms, were major controlling mechanisms for the GR runoff water quality. The seasonal variation in GR runoff water quality, according to the authors, might be due to:

- The plant-mediated nutrient cycling being altered during the growing season, either through direct plant uptake, or by the release of soil exudates;

- The chemical dissolution of minerals or ion desorption in the GR substrate, that are more rapid at higher temperatures;

- The increase of the microbial mineralization rates of the substrate organic matter with temperature, releasing dissolved inorganic nitrogen and phosphorous and solubilizing dissolved oxygen carbon;

- The season variability of GR hydrodynamics: evapotranspiration rates and rainfall event size and duration change with the season. Rapid drying occurs in the summer because of higher evapotranspiration rates, and short-duration high-intensity thunderstorms are more common in the warmer months, leading to a more rapid, episodic flow-through dynamic than in winter.

Zhang et al. [38] examined the effect on the quality of harvested rainwater of conventional roofing materials (concrete, asphalt and ceramic tile roofs) compared with a 15 cm substrate and plants GR and concluded that the ceramic tile roof was the most suitable for rainwater-harvesting applications because of the lower concentrations of leachable pollutants. The water quality of the roof runoff was closely related to the roofing material and the season when rainwater is harvested. Seasonal trends were verified in the water quality parameters of this study, which showed lower pollutants in roof runoff in summer and autumn, than those in winter and spring. The main pollutants of roof runoff from the four roofs studied, which exceeded the drinking water standard, were TN and COD. The effect of conventional roofing materials (asphalt fiberglass shingle, Galvalume-metal, and concrete tile) and alternative ones (cool and green) on harvested rainwater quality was also studied by Mendez et al. [39]. They demonstrated that if the consumer wanted to meet primary and secondary drinking water standards or non-potable water reuse guidelines, the rainwater harvested from any of these roofing materials would require treatment. The shingle roof and the GR drained water with very high concentrations of dissolved organic carbon, which might result in high concentrations of disinfection by-products after chlorination. Moreover, based on the concentrations of some metals (e.g., arsenic) in the harvested rainwater, the authors stated that the quality of the commercial growing medium should be carefully examined when the harvested rainwater is considered for domestic use.

As referred previously, the dual-substrate-layer extensive GR presented by Wang et al. [17] used porous inert materials (activated charcoal, zeolite, pumice, lava, vermiculite and expanded perlite) as the adsorption substrate, both to retain rainwater and to reduce pollutant leaching. Its performance, regarding runoff quality was assessed with six pilot-scale units (with different adsorption substrate) compared to a traditional single-substrate-layer extensive green roof. Results showed that dual-substrate-layer GR supported better natural vegetation growth, with coverage exceeding 90%, while the coverage in single substrate-layer GR was over 80%. All pilot-scale units acted as sinks for organics, heavy metals and all forms of nitrogen in all cases. They also showed that the dual-substrate-layer extensive GR performed better than the single-substrate-layer GR in retaining nitrogen, phosphorus, organics and turbidity.

When GR are enlarged to rooftop vegetable farms, the use of fertilizers might be a cause of concern. The example presented by Harada et al. [40] had atmospheric deposition of Pb and Mn exceeding the drainage output, which indicates that it is a net sink for these metals. Whittinghill et al. [14] also studied the impact of rooftop farming on stormwater runoff comparing a suite of extensive, sedum GR, located in New York City. Results indicate that the pH of runoff from the rooftop farm was slightly lower than that of the extensive GR, but the EC, apparent color, and TSS concentrations were higher, as well as the concentrations of nitrate-N, phosphorus, potassium, calcium, and magnesium. However, changes in nutrient management practices would help reduce these values.

The possible negative effects of fertilization o stormwater runoff have to be taken seriously in order not to damage the positive environmental image of GR. Emilsson et al. [41] investigated nutrient runoff, substrate nutrient storage and plant uptake following fertilization of vegetation mats, shoot-established vegetation systems and unvegetated substrate using three levels of fertilizer applied as either controlled-release fertilizer (CRF), or as a combination of CRF and conventional fertilizer. Their study clearly shows that conventional fertilizers should be avoided unless the water is either recycled or reused on the roofs or other vegetated surfaces in the first weeks following fertilization. Fertilization of old vegetation mats reduced the risk for nutrient leaching compared to fertilization of newly established roofs. They associated these results with temporary storage in the substrate and increased uptake by vegetation, however, the temporary storage of nutrients following fertilization indicated that there might be a risk for prolonged leaching. The aspects that influence the establishment of GR must be considered when designing a GR substrate and the vegetation cover. The maintenance requirements depend on the defined characteristics. Flowering and lush vegetation are not the most adequate types of plants, and fertilization might not be needed at all [41].

In summary, the runoff quality of GR is influenced by inherent and external factors of the GR, as presented in **Table 3**.

In terms of the potential for recycling of outflow water from GR, national and international standards indicate that such water can be reused for

	Pollutant source	Pollutant sinks
Inherent factors	The substrate, including component materials and depth, may contain heavy metals and nutrients that could contribute to leachate. Structural layers may contribute to heavy metal accumulation due to metal, plastic, and polymer materials used in construction. Plants are ambiguous. On one hand, they can act as an uptake of contaminants, on the other, they can be a pollutant source.	Plants behave as an uptake of contaminants by physiological metabolism of plant tissue and rhizospheric microorganisms Natural or artificial substrate added components that have the capacity of nutrient retention.
External factors	Irrigation, if supply water is contaminated (especially reclaimed water), with diverse nutrients such as N and P. Fertilizer and pesticides, including organic and chemical fertilizers and organic phosphor, chlorine, and nitrogen. Atmospheric deposition, including dry and wet deposition resulting from gravity and rainfall respectively. Age, as over time the plant biomass and ecological functions of the GR as well as levels of metal pollutants will increase, and substrate nutrients will decrease	The age of GR, can influence water conductivity and substrate pollutant retention. The age of the vegetation mat will affect the nutrient content of runoff. Newly-built GR have temporarily high nutrient levels that are steadily assimilated and degraded over time.

Table 3.
Potential of different factors, to be source or sink of pollutants (adapted from Vijayaraghavan, et al. [12]).

urban landscape irrigation and non-potable purposes such as toilet flushing. Nevertheless, treatment is required if the consumer wants to meet primary and secondary drinking water standards or demands requirement for non-potable water reuse guidelines [33, 35, 38, 39].

4. Final considerations

Stormwater control is critical to the continuous development and sustainability of urban areas, once the soil sealing caused by the increase of urbanization, together with the occurrence of more frequent extreme events due to climate change, will keep worsen the consequences of urban flood and the degeneration of water resources. GR are a type or NbS with several benefits that can help to minimize these problems.

Studies as the ones made by Karteris et al. [22] and by Hoeben and Posh [20] show the potential of implementing GR in cites (in new or existing buildings), making them more environmentally friendly, even with lack of available areas at the ground level. The majority of roof tops are unused spaces that might be transformed in a green space and a new ecosystem. The benefits of such transformation are well documented in the previous subchapters, being the most relevant the high potential of reducing CO_2 emissions (both due to the energy conservation and the CO_2

absorption), the increase of the rainwater retention rate and the delay of peak flow, which can reduce flood occurrence and negative environmental consequences to cities and its population.

Values resulting from the presented publications show GR retention rates of rainwater above 50%, reaching 100% in many cases, specifically when small-intensity rainfall events occur. Many factors were studied to understand the retention performance of GR, but it seems that the more relevant ones are the depth of substrates (intensive GR have higher retention rates than extensive ones) and the rainfalls characteristics. The retain capacity of the GR are limited and once the substrate becomes saturated, stormwater retention volume decreases significantly. However, even in those worst cases, the GR plants absorb part of the infiltrated water and extend the path taken by the water, leading to the decrease of the amount of runoff and the delay of the flood peak. Future investigations will continue to develop and search for more hydraulic-efficient GR and to understand the real effects of the large-scale implementation of GR. Long-term assessments must be made to consider variable weather conditions between years and seasons. There have been several attempts to create mathematical models to predict the runoff coefficient for distinct configurations of GR [16, 28], but the complex characteristics of these living systems, with specific dynamics and very dependent on the climatic conditions, make generalizations difficult.

Another important aspect that has been attracting the attention of investigators is the changes that GR cause on stormwater runoff quality. Plants, substrates, soil insects and microorganisms are expected to remove pollutants from rainwater in GR systems. However, the majority of the presented studies reveal that GR are a sink for heavy metals and all forms of nitrogen, but are generally a source of phosphate and dissolved inorganic and organic carbon in the growing seasons, as presented by Carpenter et al. [4], because the concentration of these parameters is higher in the runoff than in the precipitation water. This reveals significant concern about the effects of GR in urban water resources. Higher concentrations of pollutants were found in the deeper substrates and the performance of the extensive GR is, in most cases, better than the intensive systems. Intensive GR usually need fertilizers that remain in the substrates and are continuously leaching in the runoff. Since nutrients are essential for the plants especially in the first establishment years and in the growing seasons, it is not a surprise that the age of the GR has a beneficial effect in the chemical leaching of GR, as presented by the long-term studies [16, 36]. However, once that the runoff volume from GR is smaller than runoff from traditional roofs, the total discharged loads of these pollutants are smaller too. Most of the referred authors compared the GR runoff to the freshwater standards and concluded that, in general, it presents good quality and meets those standards, being also suitable for some non-potable uses. Another beneficial effect of GR reported by some authors [29, 32] is that acid deposition is neutralized by vegetation and the growing medium, revealing the potential of GR to mitigate acid rain runoff in densely populated urban areas.

Each GR is a unique system, thus differing from each other. Their nature-based characteristics make them develop and interact with the surrounding environment on their terms. The leaching problem is already defined, and solutions will be developed to minimize its consequences. First, the use of fertilizers must be very controlled and mostly avoided. Then, the implementation of layers that retain nutrients is also being considered. Also, the control of the discharged GR water may avoid the contamination of watercourses: first-flush systems might be considered, or drainage pipes can lead to gardens and/or to infiltration trenches to dispose of the nutrients in the soil. In that scope, combined solutions of GR with other LID structures must be considered when implementing GR in a neighborhood or site scale.

To pursuit circular economies, the adequacy of recycled materials in the GR substrate is also being studied which require new approaches in the study of the GR runoff quality and quantity. The use of alternative sources of water (such as greywater from the building) to irrigate might also be considered but their impacts on the runoff should be carefully assessed. Innovations will also focus on customizable low-cost and innovative GR designs, increasing the number of possible configurations, and hopefully leading to multi-perspective assessments.

Author details

Cristina Santos[1,2*] and Cristina M. Monteiro[3]

1 Faculdade de Engenharia da Universidade do Porto, Porto, Portugal

2 CIIMAR - Centro Interdisciplinar de Investigação Marinha e Ambiental, Matosinhos, Portugal

3 CBQF - Centro de Biotecnologia e Química Fina – Laboratório Associado, Universidade Católica Portuguesa, Escola Superior de Biotecnologia, Porto, Portugal

*Address all correspondence to: csantos@fe.up.pt

References

[1] EEA. Water resources across Europe — confronting water stress: An updated assessment. European Environmental Agency Report nº 12/2021. 2021

[2] FEMA. Building Community Resilience with Nature based Solutions - A Guide for Local Communities. Washington D.C., USA: Risk MAP - Increasing Resilience Together; 2021

[3] EC. Communication from the Commission to the European Parliament, the Council, the European Economic and Social Committee and the Committee of the Regions. In: Forging a Climate-Resilient Europe - The New EU Strategy on Adaptation to Climate. Brussels: European Commission; 2021

[4] Carpenter C, Todorov D, Driscoll C, Montesdeoca EM. Water quantity and quality response of a green roof to storm events: Experimental and monitoring observations. Environmental Pollution. 2016;**218**:664-672

[5] Koc K, Ekmekcioglu O, Ozger EM. An integrated framework for the comprehensive evaluation of low impact development strategies. Journal of Environmental Management. 2021;**294**:113023

[6] Perales-Momparler S, Andrés-Doménech I, Hernández-Crespo C, Vallés-Morán F, Martín M, Escuder-Bueno I, et al. The role of monitoring sustainable drainage systems for promoting transition towards regenerative urban built environments: a case study in the Valencian region, Spain. Journal of Cleaner Production. 2017;**163**:S113-S124

[7] Berndtsson J. Green roof performance towards management of runoff water quantity and quality: A review. Ecological Engineering. 2010;**36**:351-360

[8] greenroofs.org, "Green Roofs for Healthy Cities," 2021. [Online]. Available: https://greenroofs.org/about-green-roofs. [Acedido em 30 11 2021]

[9] Chen CF. Performance evaluation and development strategies for green roofs in Taiwan: A review. Ecological Engineering. 2013;**52**:51-58

[10] Liu W, Wei W, Chen W, Deo C, Si J, Xi H, et al. The impacts of substrate and vegetation on stormwater runoff quality from extensive green roofs. Journal of Hydrology. 2019;**576**:575-582

[11] Kuoppamaki K. Vegetated roofs for managing stormwater quantity in cold climate. Ecological Engineering. 2021;**171**:106388

[12] Vijayaraghavan K, Reddy D, Yun YS. Improving the quality of runoff from green roofs through synergistic biosorption and phytoremediation techniques: A review. Sustainable Cities and Society. 2019;**46**:101381

[13] Kuoppamäkia K, Lehvävirta S. Mitigating nutrient leaching from green roofs with biochar. Landscape and Urban Planning. 2016;**152**:39-48

[14] Whittinghill L, Hsueh D, Culligan P, Plunz R. Stormwater performance of a full scale rooftop farm: Runoff water qualityLeigh. Ecological Engineering. 2016;**91**:195-206

[15] Speak A, Rothwell J, Lindley S, Smith C. Rainwater runoff retention on an aged intensive green roof. Science of the Total Environment. 2013;**461-462**:28-38

[16] Akther M, He J, Chu A, van Duin B. Chemical leaching behaviour of a full-scale green roof in a cold and semiarid climate. Ecological Engineering. 2020;**147**:105768

[17] Wang X, Tian Y, Zhao X. The influence of dual-substrate-layer extensive green roofs on rainwater runoff quantity and quality. Science of the Total Environment. 2017;**592**:465-476

[18] Cristiano E, Deidda R, Viola F. The role of green roofs in urbanWater-Energy-Food-Ecosystem nexus: A review. Science of the Total Environment. 2021;**756**:143876

[19] Xu L, Yang S, Zhang Y, Jin Z, Huang X, Bei K, et al. A hydroponic green roof system for rainwater collection and greywater treatment. Journal of Cleaner Production. 2020;**261**:121132

[20] Hoeben A, Posch A. Green roof ecosystem services in various urban development types: A case study in Graz, Austria. Urban Forestry & Urban Greening. 2021;**62**:127167

[21] Hashemi S, Mahmud H, Ashraf M. Performance of green roofs with respect to water quality and reduction of energy consumption in tropics: A review. Renewable and Sustainable Energy Reviews. 2015;**52**:669-679

[22] Karteris M, Theodoridou I, Mallinis G, Tsiros E, Karteris A. Towards a green sustainable strategy for Mediterranean cities: Assessing the benefits of large-scale green roofs implementation in Thessaloniki, Northern Greece, using environmental modelling, GIS and very high spatial resolution remote sensing data. Renewable and Sustainable Energy Reviews. 2016;**58**:510-525

[23] Grunwald L, Heusinger J, Weber S. A GIS-based mapping methodology of urban green roof ecosystem services applied to a Central European city. Urban Forestry & Urban Greening. 2017;**22**:54-63

[24] Rowe D. Green roofs as a means of pollution abatement. Environmental Pollution. 2011;**159**:2100-2110

[25] Liu H, Kong F, Yin H, Middel A, Zheng X, Huang J, et al. Impacts of green roofs on water, temperature, and air quality: A bibliometric review. Building and Environment. 2021;**196**:107794

[26] Harper G, Limmer M, Showalter W, Burken J. Nine-month evaluation of runoff quality and quantity from an experiential green roof in Missouri, USA. Ecological Engineering. 2015;**78**:127-133

[27] Lee J, Lee M, Han M. A pilot study to evaluate runoff quantity from green roofs. Journal of Environmental Management. 2015;**152**:171-176

[28] Viola F, Hellies M, Deidda R. Retention performance of green roofs in representative climates worlwide. Journal of Hydrology. 2017;**553**:763-772

[29] Zhang Q, Miao L, Wang X, Dandan L, Zhu L, Zhou B, et al. The capacity of greening roof to reduce stormwater runoff and pollution. Landscape and Urban Planning. 2015;**144**:142-150

[30] Liu W, Engel B, Chen W, Wei W, Wang Y, Feng Q. Quantifying the contributions of structural factors on runoff water quality from green roofs and optimizing assembled combinations using Taguchi method. Journal of Hydrology. 2021;**593**:125864

[31] Gregoire B, Clausen J. Effect of a modular extensive green roof on

stormwater runoff and water quality. Ecological Engineering. 2011;**37**:963-969

[32] Berndtsson J, Bengtsson L, Jinno K. Runoff water quality from intensive and extensive vegetated roofs. Ecological Engineering. 2009;**35**:369-380

[33] Beecham S, Razzaghmanesh M. Water quality and quantity investigation of green roofs in a dry climate. Water Research. 2015;**70**:370e384

[34] Akther M, He J, Chu A, van Duin B. Nutrient leaching behavior of green roofs: Laboratory and field investigations. Science of the Total Environment. 2021;**754**:141841

[35] Razzaghmanesh M, Beechama S, Kazemi F. Impact of green roofs on stormwater quality in a South Australian. Science of the Total Environment. 2014;**470-471**:651-659

[36] Todorov D, Driscoll C, Todorova S, Montesdeoca M. Water quality function of an extensive vegetated roof. Science of the Total Environment. 2018;**625**: 928-939

[37] Buffam I, Mitchella M, Durtsch R. Environmental drivers of seasonal variation in green roof runoff water quality. Ecological Engineering. 2016;**91**:506-514

[38] Zhang Q, Wang X, Hou P, Wan W, Li R, Ren Y, et al. Quality and seasonal variation of rainwater harvested from concrete, asphalt, ceramic tile and green roofs in Chongqing, China. Journal of Environmental Management. 2014;**132**:178-187

[39] Mendez C, Klenzendorf J, Afshar B, Simmons M, Barrett M, Kinney K, et al. The effect of roofing material on the quality of harvested. Water Research. 2011;**45**:2049-2059

[40] Harada Y, Whitlow T, Russell-Anelli J, Walter M, Bassuk N, Rutzke M. The heavy metal budget of an urban rooftop farm. Science of the Total Environment. 2019;**660**:115-125

[41] Emilsson T, Berndtsson J, Mattssona J, Rolfa K. Effect of using conventional and controlled release fertiliser on nutrient runoff from various vegetated roof systems. Ecological Engineering. 2007;**29**:260-271

Chapter 9

Nature-Based Solutions Using LECA LWA to Increase Urban Sustainability and Support Stormwater Management

Cristina M. Monteiro, Cristina Santos, Jaran R. Wood and Kim Rosenbom

Abstract

In recent decades, the increase of building systems and the consequent impermeabilization of the inner-city environment, poses several negative environmental risks, especially those regarding global warming and water management. Urban green infrastructure (GI) provides benefits to urban inhabitants, but their potential depends on their suitable implementation, under the responsibility of different stakeholders. Nature-based solutions (NbS) are an effective means to restore vegetation urban areas contributing to its sustainability and resilience. The use of NbS infrastructure helps to reduce flood incidences, furthermore contributing to the mitigation of the urban heat island effect and air pollution reduction. Several NbS such as green roofs and green walls, rain gardens, infiltration trenches and permeable pavements, are being implemented worldwide as an effective strategy to improve the environmental performance of densely populated urban cities. Lightweight Expanded Clay Aggregate (LECA®) has been widely incorporated in several innovative urban GI solutions due to their key characteristics (e.g. lightweight, porosity and good drainage), developing a crucial role in stormwater management and water runoff quality. This chapter aims to report the potential influence that several NbS have on urban stormwater management, considering several case studies using Leca® LWA—demonstrating their engineering benefits and innovative properties.

Keywords: nature-based solutions, urban green infrastructures, urban stormwater management, urban sustainability, lightweight expanded clay aggregate (LECA®), climate resilience

1. Introduction

The European Commission [1] states that collectively our people, our planet and our prosperity are vulnerable to climate change, and points out the need to prevent the un-adaptable and to adapt to the un-preventable. The increase in the frequency and severity of climate and extreme weather conditions has created a surge in the number of, and damages from, environmental disasters in recent decades. The implementation of the European Union (EU) Strategy on the Adaptation to Climate Change intends to put its countries in a better position to face climate impacts by 2030. This would mean (1) adaptation awareness and changes to every single local authority, company and household; (2) adaptation implementation is to be well underway for those most affected; and (3) global leadership is to immediately commence in areas such as climate services, climate proofing, or nature-based solutions. As climate change becomes more intense, the urgency of adaptive measures increases. The Strategy outlines a long-term vision for the EU to become a climate-resilient society and aims to reinforce the adaptive capacity of its countries and the rest of the world—thus minimising our vulnerability to the impacts of climate change, such as droughts, extreme weather events and flooding [2]. In this Strategy, ecosystem-based approaches are presented as a cross cutting priority. They focus on ecosystem restoration and on the enhancement of ecosystem services, being recognised as multipurpose solutions that are often more efficient than traditional technical measures.

Climate change adaptation and disaster risk reduction through the implementation of ecosystem-based approaches are widely applicable and hugely varied. They encompass several related concepts, such as Nature-based Solutions (NbS) and Green Infrastructure, that aim to enhance social and environmental resilience by restoring, maintaining, and improving ecosystems—thus enhancing their services to society, in areas such as water retention and prevention of soil erosion, floods and droughts. They answer to several environmental and sectoral policy objectives (e.g. regarding biodiversity, water quality or agricultural and forest management) and generate multiple socio-economic benefits that often go beyond technical solutions [2].

Nature-based Solutions are defined as solutions to societal challenges that are inspired and supported by nature, which are cost-effective, simultaneously provide environmental, social and economic benefits and help to build resilience [3]. Such solutions bring nature and natural systems and processes into cities, landscapes and seascapes, through locally adapted, resource-efficient and systemic interventions. Despite the use of different terms referring to NbS used in international and European policies, the fundamental overarching idea is that nature can be used as a valuable tool to strengthen the resilience of ecosystems, protect biodiversity and reduce the risk of extreme weather and climate related disasters. Furthermore, NbS can help address broader societal challenges, including social and economic challenges within the paradigm of sustainable development.

For the effective design and successful implementation of NbS, it is critical to ensure that the measures are adequate for addressing the hazards while also delivering other societal benefits. If well designed, NbS can provide sustainable, cost-effective and multipurpose possibilities that can act as alternatives to or be applied to create synergy with established grey infrastructure [3]. Their benefits cover a large range of sectors, but specifically for water management and urban areas, the main ones are presented in **Figure 1**.

Nature-based solutions are essential for sustaining clean and healthy water resources. They help to address the risk of flooding and water scarcity (drought) by regulating water flows. Using NbS inland, including the restoration of the sponge-like function of soils, will boost the supply of clean and fresh water and furthermore reduce the risk of flooding. On the other hand, increasing green spaces can increase water infiltration into the soil, enhance evapotranspiration and provide storage areas for rainwater, which can alter the magnitude and timing of water runoff and flooding

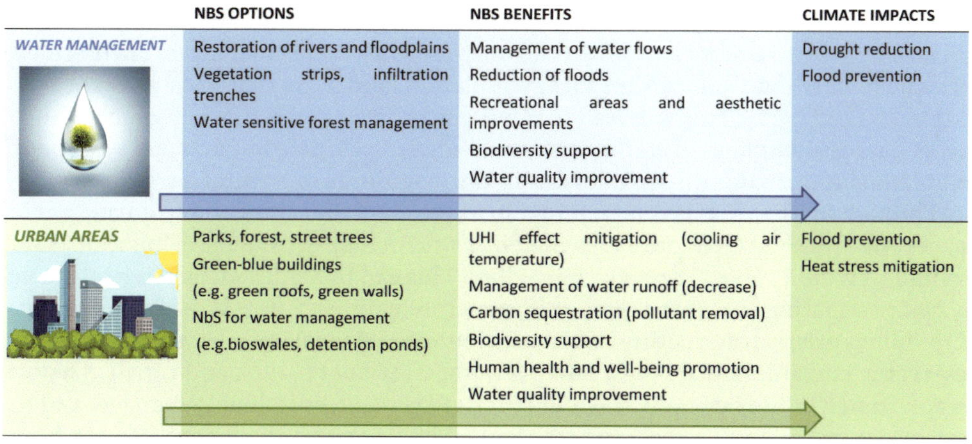

Figure 1.
Multiple benefits of nature-based solutions for addressing climate hazards in urban areas (adapted from EEA report n°01/2021 [3]).

during heavy precipitation events, while contributing to maintaining water flow during drought periods. Moreover, creating new green space for stormwater management (e.g. constructed wetlands, bioswales) has been shown to enhance flood protection while providing additional benefits such as biodiversity protection, recreational green urban space and water purification.

The porosity and hydraulic conductivity of lightweight aggregates (LWA) materials fulfil the hydraulic performances required for their use within hydraulic urban infrastructures. The application of LWA in urban drainage systems may guarantee an infiltration rate adequate for reducing different forms of stormwater runoff (e.g. rain gardens, infiltration basins, infiltration trenches). In addition, the sorbent capacities demonstrated by LWA complement their features in promoting both the infiltration of urban runoff stormwater and the recharging of groundwater, meeting the established quality requirements [4]. LECA® is the brand name that manufactures the LWA product, which is, an heterogeneous group of low-density material used for various civil engineering and construction applications [5]. LECA® LWA has been increasingly applied in stormwater management schemes and urban green infrastructure including green roofs and walls, permeable pavements and thermal insulation concretes [6–10].

This chapter aims to explore the application of LECA® LWA in NbS and to demonstrate how its application increases performance in terms of water management within urban areas. As discussed, NbS such as green roofs and green walls, rain gardens, infiltration trenches and permeable pavements, are being implemented worldwide as a solution to improve the environmental performance of densely populated urban cities. This performance can be enhanced by incorporating LECA® LWA, due to its natural characteristics (e.g. lightweight, porosity and good drainage) as presented in the following sections.

2. Urban nature-based solutions

Cities worldwide are facing challenges of resilience, as climate risks interact with urbanisation, loss of biodiversity and ecosystem services, poverty, and rising socioeconomic inequality [11]. Extreme events, especially flooding, heat waves and droughts, are causing significant economic losses and affecting the wellbeing of citizens and the rise of urbanisation will only magnify these challenges. Furthermore, cities host more than half of the global population, and more than 70% are expected to do so by 2050 [11]. This will lead to rapidly urbanised areas, most of which with lower-quality—unplanned settlements, vulnerable to climate impacts and with poorly maintained infrastructure—vulnerable to the magnitude of natural hazards.

Through this basis, it is fundamental to build adaptable cities and to create new spaces with resilient characteristics that, in a situation of extreme precipitation, for example, create a robust regeneration ability. This can be achieved, as referred above, by complementing traditional grey infrastructure with NbS. The World Bank group [11] defines grey infrastructure as built structures and mechanical equipment, such as reservoirs, embankments, pipes, pumps, water treatment plants, and canals. They can be considered as the opposite of NbS (actions to protect, sustainably manage, and to restore natural or modified ecosystems that address societal challenges effectively and adaptively, simultaneously providing human well-being and biodiversity benefits, as presented) but, in fact, they are almost in-dissociable and function in a complementary way. The following subchapters present different types of NbS that can be implemented in urban areas.

2.1 Neighbourhood/site scale NbS

Nature-based solutions promote resilience and adaptation while being integrated into a community's-built environment or in its natural surrounding areas [12]. Nature-based stormwater systems, wetlands and green roofs, for example, can alleviate flooding from intense rainfall events, and also improve water and air quality, whilst reducing urban heat-island effects and creating ecological corridors for biodiversity. They are an effective means to restore vegetation in densely populated areas.

NbS can also be considered as environmentally-conscious approaches to stormwater management in urban areas, also known as Sustainable urban drainage systems (SUDS), low impact development (LID) or best management practices (BMP's) [12]. They help to control the stormwater runoff but also create a positive impact on its quality, minimising the effects of untreated stormwater, traditionally drained by grey infrastructure, on receiving water bodies. On the other hand, the implementation of NbS within urban areas, will promote the infiltration on stormwater, which will reduce the amount of water drained to Wastewater Treatment Plants (sewage networks, even when separative, have rainfall derived inflow and infiltrations that lead to the excessive entrance of rainwater into WWTP), resulting in important economic gains.

In conclusion, NbS can support surface water and groundwater water protection, and contribute to the qualitative and quantitative environmental and ecological status of water bodies. In a large scale, the extension of buffer zones and the establishment of water-sensitive forests and NbS for urban areas will reduce pollution and support habitat quality and biodiversity conservation, which can have positive effects for developing businesses and jobs, in particular those related to recreation and tourism. However, there are also some challenges related to the implementation of NbS, such as: (1) the associated construction/development works, which may negatively affect water quality and river ecosystems in the short-term; (2) enhanced green spaces connectivity triggers the dispersal of unwanted organisms (e.g. mosquitoes) with negative impacts on both local ecosystems and human health; (3) conflicts between different sectors, for example if agricultural areas are temporarily used as reservoirs for flood expansion; (4) in the case of heavily polluted surface waters, the reconnection of floodplains with rivers can contribute to the wide-scale diffusion of pollutants to soils, agricultural areas and groundwater [3]. Careful planning with a large-scale comprehensive analysis is thus essential when implementing a more natural and resilient configuration of cities. To initiate NbS, two main questions are fundamental: "What benefits can be obtained from the desired NbS?" and "Is it adequately suitable for the implementation onto site?"

According to FEMA [12], the implementation strategies for NbS are diverse, and a one size fits all strategy is not suitable. Choosing a solution depends on a several factors, including the level of natural hazard risk reduction, land use planning, economics and more [12]. This agency classifies NbS in three main categories, based on scale and location:

- Watershed or landscape scale: interconnected systems of natural areas and open space (large-scale practices that require long-term planning and coordination);

- Neighbourhood or site scale: distributed stormwater management practices that manage rainwater where it falls (they can often be built within a site, corridor, or neighbourhood without requiring additional space);

- Coastal areas: NbS that stabilise the shoreline, reducing erosion and buffering the coast from storm impacts. While many watershed and neighbourhood-scale solutions work in coastal areas, these systems are designed to support coastal resilience.

Watershed-scale NbS include land conservation (preserving interconnected systems of open space that sustain healthy communities), greenways (corridors of protected open space managed for both conservation and recreation), wetland restoration and protection, stormwater parks (recreational spaces designed to flood during extreme events and to effectively manage stormwater), floodplain restoration (keeping waterways healthy by storing floodwaters, reducing erosion, filtering water pollution, and creating a natural habitat).

Coastal-areas NbS include coastal wetlands, oyster reefs, dunes, waterfront parks (intentionally designed to flood during extreme events, reducing flooding elsewhere), living shorelines (combination of living components, such as plants, with structural elements, such as rock or sand).

The neighbourhood or site specific NbS are mainly found within cities and urban spaces. The most common ones are presented in the following sections.

2.1.1 Green roofs

A Green Roof (GR—also known as living roof or vegetative roof) is a vegetated system installed on a constructed roof structure consisting of several materials settled in a multilayer composition which must provide a suitable breeding ground for vegetation, respecting and promoting the physical integrity of the built structure (**Figure 2**). The multilayer system consists of the following layers (from top to bottom): a growing substrate layer, a separation filter, a drainage layer, a protection and absorption filter and a root barrier.

An effective GR system can absorb stormwater and temporarily store it to be used by the vegetation in their biological processes, reducing the amount of water running

Figure 2.
Example of Green roof [13].

off into the stormwater drainage system and furthermore improve its quality. A GR is a highly appreciated water management solution through its ability to decrease stress on the sewer systems at peak flow in urban areas. According to Augustenborg Botanical Roof Garden (Malmo, Sweden) research paper, a GR retains 50% of the rainwater on yearly basis, allowing only for the 50% of the rainwater to runoff into the drainage system. However, the retention capacity of a GR is highly dependent on the climate conditions, so a significant variability on the runoff retention capacity can be expected throughout a year in different locations.

2.1.2 Vegetated swales

Vegetated swales are an open channel system holding plants or mulch to treat and infiltrate stormwater (**Figure 3**). Grassed channels, dry and wet swales and biofilters are included in this type of NbS. Though there are differences in their design, they are all an upgraded form of a traditional drainage ditch. According to Jotte et al. [13], the swale technique cannot be applied to large drainage areas, but their small slope makes them a good choice for treating runoff from highway or residential areas.

Vegetated swales are designed to slow runoff, promote infiltration, and filter pollutants and sediments in the process of the transmission of water runoff, through the use of endemic vegetation species, capable of growing in water environments and with the capacity to remove pollutants [15]. Pollutants are removed by sedimentation and the filtering processes, both promoted by vegetation and by the different layers of the soil.

2.1.3 Soakaways

Soakaways are small infiltration devices built close to the receiving stormwater production areas. Their walls and bottom are permeable and allow water to flow into the surrounding soils (**Figure 4**), thus their main design considerations are the permeable area and the infiltration rate. The performance of these structures is highly dependent on the soil characteristics and high levels of groundwater can make this NbS inadequate.

Figure 3.
Example of vegetated swale [14].

Figure 4.
Example of a soakaway [15].

Usually, soakaways serve only one household and are constructed in the private grounds surrounding a property, allowing for rainwater management to be handled at the source. Their main benefits are related to runoff volume reduction, hydraulic control and groundwater recharge, but can furthermore through the possibility of complementing other NbS (rain barrels or other rainwater harvesting systems); they also have a very positive effect on stormwater quality, due to the removal of total suspended solids, heavy metals, phosphorous and nitrogen removal [13].

2.1.4 Filter/buffer strips

Unlike swales, buffer strips are set on flat areas with very low slopes. They may be located along streets and highways in order to receive the lateral runoff to block suspended solids and associated pollutants (**Figure 5**) [13].

Grass or other dense vegetation are planted in buffer strips to treat the runoff through vegetative filtering, sedimentation, and to promote water retention and infiltration when possible. Local wild grass and flower species can also be introduced for visual interest and to provide a wildlife habitat [15]. This NbS can serve as a buffer between incompatible land uses and can provide locations for groundwater recharge in areas with pervious soils.

Figure 5.
Example of a vegetated filter strip [15].

2.1.5 Infiltration trenches and basins

Infiltration trenches and infiltration basins (**Figure 6**) have the same operating principle. Similarly, to soakaways, they are designed to receive the surrounding stormwater and to infiltrate it to the surrounding soils through the walls and the base. However, they have an open configuration which leads to the necessity of designing and creating an effective pre-treatment unit to remove coarse solids and hydrocarbons.

Groundwater contamination and soil saturation are concerns associated with these NbS, hose performance is dependent, amongst other factor, on the characteristics of the soil [13]. However, many benefits are linked to these NbS, especially related to the reduction of stormwater runoff, groundwater recharge and treatment capacity (filtration and bio filtration through the substrate in the trench and subsequently through soil).

2.1.6 Wetlands

Wetlands are artificial ponds, conceived with emerged, submerged and/or floating plants, that have a significant role in the removal of a significant part of the influent pollutants (**Figure 7**). They are made up with a substrate layer (except for floating

Figure 6.
Examples of an infiltration trench (left) and an infiltration basin (right) [15, 16].

Figure 7.
Example of an urban wetland [17].

vegetation) and water appropriate plants, being separate from the surrounding soil by an impermeable membrane. This configuration allows stormwater to be detained in the pond and benefits from the natural depuration processes. Part of the water is consumed by the plants and the remaining one leaves the wetland with its quality significantly improved.

Properly designed wetlands can remove significant amounts of nitrogen and phosphorus, suspended solids, and other pollutants from urban environments [13]. When properly designed, wetlands can also provide landscape enhancement and harvest the discharged water for non-potable purposes.

2.1.7 Raingardens

Raingardens (also known as bioretention, bioswales or biofilters) are shallow excavated surface depressions containing mulch and a prepared soil mix, planted with specially selected native vegetation that captures and treats runoff (**Figure 8**), through filtration, which then can either infiltrate to the natural soil or be collected into a sublayer returning to the stormwater sewer system. They remove pollutants from runoff through filtration in the soil and uptake by plant roots and can help to reduce runoff volume through evapotranspiration and full or partial infiltration [13].

The main function of these gardens is to receive, treat and infiltrate low-intensity but frequent precipitation, so they can be complemented with other NbS or even with

Figure 8.
Example of a rain garden [15].

grey-infrastructure to manage excess flows. The use of native vegetation is essential to keep a well maintained and sustainable structure.

2.1.8 Permeable pavements

Pavements with high infiltration capacity are considered permeable. Nowadays, considering the significant area that streets and highways encompass in urban spaces, the transformation of such surfaces into permeable ones are fundamental to reduce the negative impacts of soil sealing. Infiltrating water can thus flow directly into the underlying layers (soils and aquifers) or be stored below ground and released at a controlled rate to surface water.

Examples of permeable pavements are permeable asphalt, permeable concrete, permeable interlocking concrete pavers, concrete grid pavers (**Figure 9**), and plastic grid pavers where opening are filled with pea gravel, sand or top soil and grass [13].

Figure 9.
Example of a permeable pavement [18].

Urban NBS	Advantages	Disadvantages
Green Roofs	Stormwater runoff retention and large detention capacity. Aesthetically pleasing. Good insulation properties. Reduction in impervious area of the property. Suitable for high density developments.	Maintenance of roof vegetation. Expensive design and construction when compared to traditional tile roofs. Necessity of drainage systems and, in some cases, irrigation also. Limited plant species. Potential fire hazard.
Vegetated Swales	Less expensive than conventional conveyance practices. Enhances infiltration. Reduces runoff rates and volumes.	Water treatment is limited to a confined area. There are risks of blockage in connecting pipe networks.
Soakaways	Minimal area space is required. Easy construction and operation. Groundwater recharge.	Not suitable for poor drainage soils. Not suitable for polluted runoff. Performance is reduced during wet periods.
Infiltration trenches	Significant reduction of runoff rates and volumes. Effective reduction of many pollutants, including suspended solids, bacteria and trace metals.	High clogging potential without effective pre-treatment.

Urban NBS	Advantages	Disadvantages
Infiltration basin	Reduction of peak flow rate, erosion and scouring. Effective pollutant removal and local flooding reduction. Groundwater recharge. Base flow of nearby streams is maintained. Simple and cost-effective solution.	High potential for failure rates. Potential for pollution if high TSS (solid materials, including organic and inorganic, that are suspended in the water) in the runoff. Not appropriate for large drainage areas (above 10 acres).
Wetlands	Effective pollutant removal. Enhance wildlife. Good aesthetic and high ecological potential. Positive community acceptance. May increase value of property.	Need to be lined and applied where there is a risk of groundwater contamination. Not appropriate for steep groundwork sites. Nutrients release. No significant runoff volume reduction Space restrictions. May require maintenance and approval from safety authorities. Risk of invasive species development.
Rain gardens	Aesthetic enhancement. Minimal land consumption. Runoff rates and volumes reduction.	Risk of clogging if there is a poor maintenance of the surrounding landscape. Requires proper plant selection and maintenance. Treats relatively small drainage areas.
Permeable pavements	Significant reduction of the runoff rates and volumes. Impervious surface area reduction. Provides pollutant filtering.	Not suitable for high sediment loads.

Table 1.
Advantages and disadvantages of source control systems (adapted from [13]).

2.2 Brief summary

Other NbS can also be classified (e.g. detention ponds or rainwater harvesting systems), but are considered beyond the scope of this chapter because their benefits cannot be improved by using LECA® LWA. The following table presents a comparative analysis of the urban NbS described previously. For each technique, a small description and some advantages and disadvantages are included (**Table 1**).

3. Use of LECA® LWA IN nature-based solutions

3.1 General considerations

The main advantage of NbS in relation to urban stormwater management, is the reduction of the urban surface water runoff that goes into drainage networks (and sometimes to wastewater treatment plants (WWTPs) through the sewer systems), and the delay of the peak flow, when compared to traditional systems without detention mechanisms, as seen in **Figure 10**.

However, NbS efficiency can be enhanced through the application of LECA® lightweight aggregate (LWA) in its different layers, by itself or mixed with other components. LECA® LWA in a sublayer of a NbS, increases the temporary storage

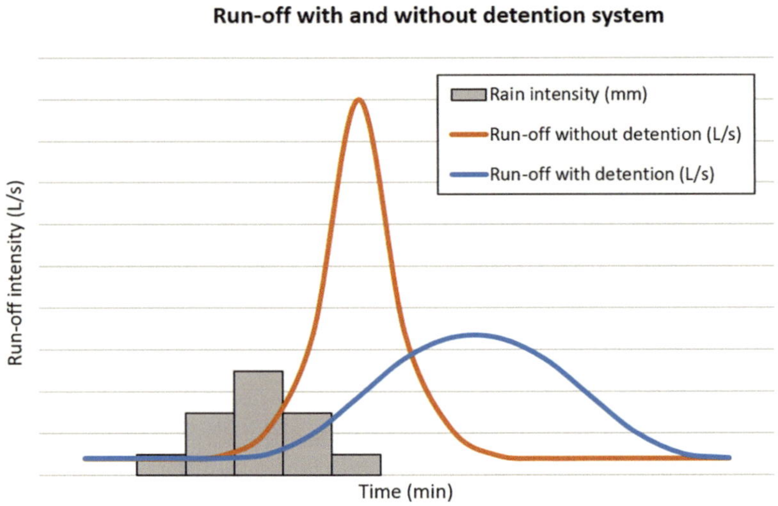

Figure 10.
Typical curve of runoff from a drainage system, with and without a detention layer [19].

capacity for water, reducing the average runoff intensity and decreasing the peak flow intensity. As such, the use of LECA® LWA maximises the advantages and functions of a NbS, as well as the local water resilience.

LECA® LWA is a natural product made by heating natural clay and can be used as a construction material in local water management infrastructure, due to its high mechanical strength, robustness, easy installation and sustainable longevity. LECA® LWA competes with other materials on the market for water management such as plastic boxes, textile mats, pumice, foam-glass and crushed brick [20]. This aggregate has an inherent ability to retain, detain and drain water runoff due to its natural characteristics: lightweight highly porous grains, high specific surface area and large volume of voids between the grains. The geometry allows LECA® LWA to temporarily detain moderate amounts of flowing water and thereby reduce peak intensity runoff for small catchments.

Furthermore, due to its highly porous structure and large surface area of crushed grains, LECA® LWA is also suitable to improve rainwater runoff quality. Through physical, biological and chemical processes, LECA® LWA can act as a filter, removing dissolved and suspended particles, nutrients (e.g. phosphates) and heavy metals in its porous grains that would otherwise end up downstream within the water cycle. Its greater resistance to clogging enhances its use in stormwater treatment and is furthermore supported by its typically low operating costs and a long life-span [19].

The capacity of a NbS to manage stormwater, is determined through the runoff coefficient or coefficient of discharge (C), a dimensionless value which varies between 0 and 1 that indicates the surface runoff from a catchment. The runoff coefficient is expressed as the rational between the total amount of water running onto the surface and the amount of received precipitation during a rainfall. **Figure 11** presents runoff coefficients for different surfaces with increasing precipitation.

The application of a subsurface layer of LECA® LWA into an already existing permeable surface (**Figure 11**) allows for lower runoff coefficients, achieved even at high rain intensities, when compared to other surfaces. Highly porous crushed LECA® LWA

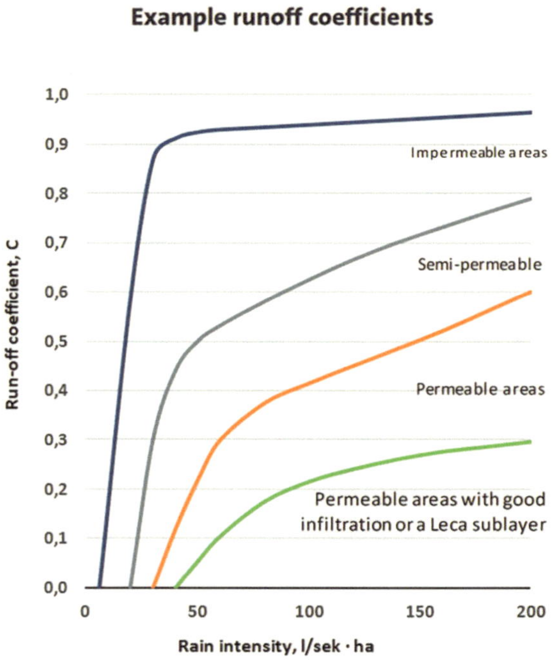

Figure 11.
Examples of runoff coefficient (C) for different surfaces with increasing rain intensity [19].

integration in a given water management solution, substantially decreases the runoff coefficient of the surface, through retention and detention of rainwater, thus improving the capacity to manage water runoff in extreme events, delaying water runoff and preventing floods. LECA® LWA will operate in synergy with the surface components, and the runoff reduction will increase with the thickness of LECA® LWA layer.

3.2 Hydrological performance of LECA® LWA with different granulometry

LECA® LWA can be presented with different granulometry: finely crushed (0–6 mm); coarsely crushed (4–10 mm) and uncrushed or round LECA® LWA (10–20 mm). In order to assess LECA® LWA hydrological characteristics, technical water retention tests simulating an intense rainfall, have been performed, in trial tests of 100 mm and 200 mm layer thickness, 2% slope and compared to a reference impermeable area without detention measures. The achieved results are shown in **Figure 12** and the corresponding values presented in **Table 2** [20].

The results have shown that finely crushed LECA® LWA in both layer thickness tests, presented higher peak runoff intensity reduction and a higher amount of detained water, when compared to the coarsely and round LECA® LWA. This result solves the gap identified by Pradhan et al. [21] when studying the hydraulic performance of several plant growth media in a greywater treatment NbS, including one with LECA® LWA. The low performance of this aggregate when comparing to perlite, coco coir and sand, was due to the largest size particles and high porosity. On the other hand, the runoff reduction resulting from the finely and the coarsely crushed LECA® LWA, and showed in **Table 2**, is with accordance with the reduction expected from an intensive green roof, as the one reported by Lee et al. [22] with 200 mm substrate layer thickness and a runoff reduction between 42.8–68%.

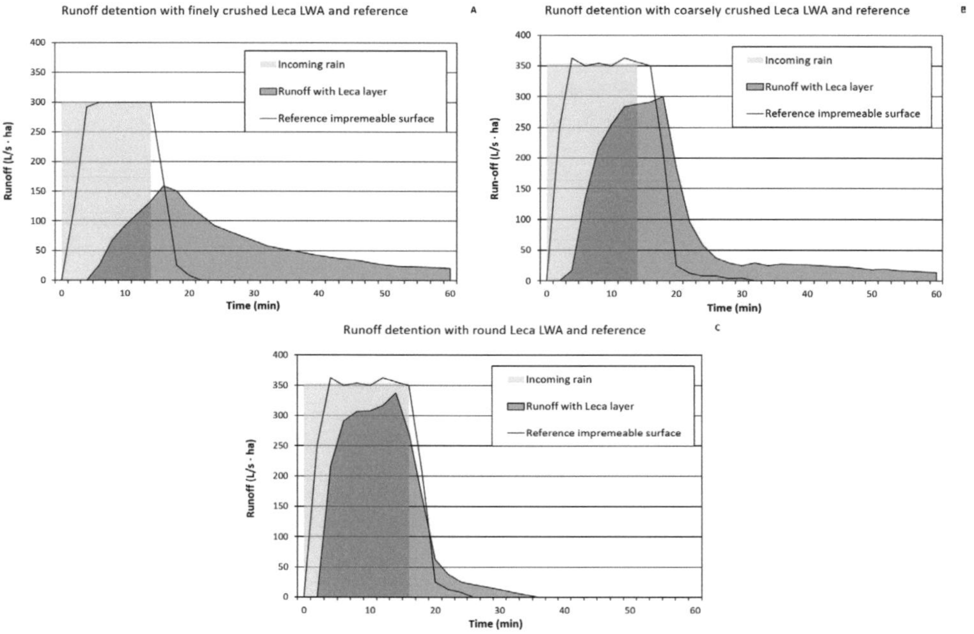

Figure 12.
Runoff behaviour for LECA® LWA (A: Finely crushed; B: Coarsely crushed; C: Round).

LECA® type	Finely crushed LECA®		Coarsely crushed LECA®		Round LECA®	
Rainfall event conditions simulation	27 L/m^2 for 15 min		36 L/m^2 for 15 min		36 L/m^2 for 15 min	
Rain intensity	300 L/(s.ha)		350 L/(s.ha)		350 L/(s.ha)	
Thickness of LECA®-layer	200 mm	100 mm	200 mm	100 mm	200 mm	100 mm
Peak runoff intensity reduction	53%	45%	35%	18%	17%	6%
Amount of delayed water, approx.	70%	55%	40%	30%	20%	15%

Table 2.
Peak runoff intensity (%) and the total amount of water detained (%) for different LECA® LWA granulometry.

Tests for runoff coefficient determination of the three LECA® LWA granulometry showed that finely crushed LECA® LWA presents the best runoff coefficients: values between 0.15–0.4 in a 200 mm thickness of LECA® LWA layer (with rain intensity varying from 200 to 400 L/(s.ha)) when compared to coarsely and round LECA® LWA with values between 0.2–0.5 and 0.3–0.7, respectively. The lower runoff coefficients achieved by finely crushed LECA® LWA are explained due to their higher surface area and porosity that can retain water [20].

	Pollutant removal (%) by filter material			
	Cu	**Pb**	**P**	**Zn**
LECA® round 2–4 mm	99	99	>90	98
LECA® crushed 0–3 mm	99	99	>90	98
LECA® crushed 3–8 mm	86	85	>90	80

Table 3.
Stormwater pollutant removal (%) in laboratory column tests (initial pollutants concentration: Cu = 10 mg/L; Pb = 20 mg/L; P = 10 mg/L; Zn = 40 mg/L).

Therefore, based on the technical tests performed and depending on the intended purpose when installing a LECA® LWA layer coupled with a NbS, it can be claimed that finely crushed LECA® LWA is optimal when maximum water detention is intended; coarsely crushed LECA® LWA when higher hydraulic permeability must be achieved; and round LECA® LWA when increased drainage and water storage are needed.

LECA® LWA can also act as a stormwater filter material—improving water runoff quality. By incorporating LECA® LWA into biofilters or as filter strips, a wide range of dissolved and suspended pollutants and heavy metals can be removed from water runoff. Holt et al. [23] reported removals of several pollutants using different granulometry of LECA® LWA as presented in the following **Table 3**, based on the results achieved on the VTT Stormfilter research program.

3.3 Runoff coefficient for different layered detention-based roofs with LECA® LWA

The work developed by Schärer [24] to investigate runoff coefficients for different layered detention-based roofs, included a comparison of different roof configurations with commonly used layers of typical combinations of green or LECA® LWA—based roof solutions. Results confirm the beneficial effect of LECA® LWA on the performance of green roofs and permeable pavements.

From the five configurations studied for single layer roofs, the 10 cm LECA® medium layer presented the best performance with a lower and later peak flow. It had a considerably lower runoff coefficient than the other four individually tested layers (two with two types of felt mat and two with drainage mat with and without holes on the bottom).

Three green roofs were also tested: one with sedum and felt mat, another with sedum, felt mat over-the-drainage mat with extra drainage holes and the last one with sedum, felt mat, drainage board and LECA® medium (**Figure 13**). This last one resulted in the lowest runoff coefficient and also gave a substantially lower runoff coefficient than the 10 cm LECA® medium layer referred above.

LECA® LWA also had a very significant effect in the performance of the studied non-vegetated roofs (**Figure 14**). The runoff coefficient of a LECA® LWA based roof system with permeable pavement (C = 0.33) was significantly lower than the test of the pavement alone (C = 0.89).

Of all the experimental measurements (ten roofs, plus one reference), the LECA® medium based systems gave the lowest runoff coefficients. Those were the most complex systems, with the main drainage in the LECA® LWA layer. The event-based detention hydraulic performance was found to be higher for vegetated roofs, instead of the non-vegetated ones. LECA® medium clearly made a significant difference on

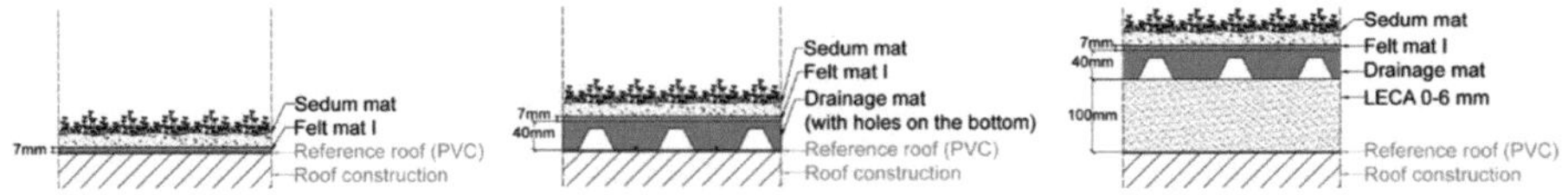

Figure 13.
Green roofs configuration [24].

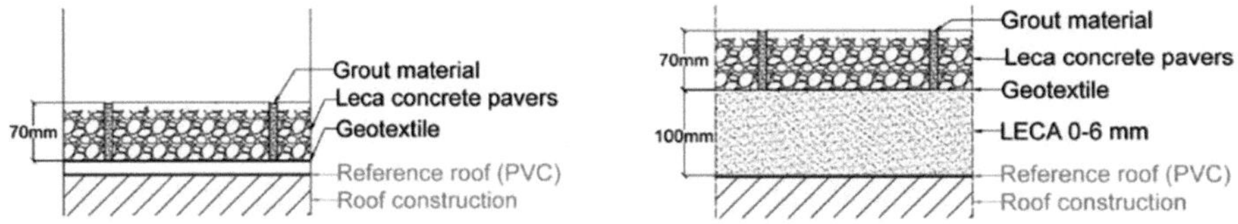

Figure 14.
Non-vegetated roofs configuration [24].

the results with lowest measured runoff coefficients on the corresponding roofs that achieved runoff coefficients of around 0.3 instead of around 0.8 in the corresponding roofs without it.

Koc et al. [25] reported the capacity of four different NbS for stormwater management presenting a runoff reduction of 41.67% for green roofs, 37.83% for bioretention cells, 27.99% for infiltration trench and 29.04% for permeable pavement. Another study reported by Harper et al. [26] presented ca. 60% runoff reduction of a vegetated substrate compared and only 40% runoff reduction when the same substrate was unplanted. Although this shows a good stormwater management performance, according to the findings of Schärer [24] here presented, these NbS could improve their runoff retention values by introducing LECA® LWA in their layers.

3.4 LECA® LWA field test at Høvringen in Trondheim and the Oslo project

For a long time, the traditional way of guiding rainwater off a roof has been with gutters and downspouts to the ground, which nowadays faces two major challenges: (1) more precipitation and more frequent short periods of torrential rain (resulting in large quantities of water overloading stormwater sewage networks); (2) densification of urban areas and fewer natural areas for absorbing and infiltrating water. As such, in 2016 under the project Klima 2050 (developed at Høvringen in Trondheim, Norway), researchers started focusing on new solutions for delaying water runoff when it rains while at the same time converting the sewage treatment plant roof into an attractive terrace. As such, a field test for detention and retention based roofs was installed and instrumented at Høvringen RA in Trondheim, Norway (**Figure 15**) [13].

The detaining and retaining runoff properties of the pilot GR system have shown that both mean and peak intensity (5 min) of runoff may be reduced. The LECA® LWA based configuration provided a detention performance for a peak runoff reduction of 95% (median) and for a peak delay of 1 h 15 min (median). The peak detention and peak reduction from reference vs. blue grey roof achieved 1 h 47 min and 75%, respectively. The high permeability characteristic of LECA® LWA contributed

Figure 15.
KLIMA 2050 project field test at Høvringen using LECA® LWA in Trondheim, Norway (field 1—LECA® LWA below permeable paving blocks; field 2—Black empty reference roof; field 3—LECA® LWA under a traditional sedum green roof) (Photo: Tore Kvande, NTNU—Norwegian University of Science and Technology).

to a steady, but restricted flow of water through the medium. Therefore, it could be concluded that LECA® LWA can be used as an alternative to natural detention and infiltration of rainwater, to reduce local runoff coefficients, by simulating the same mechanisms found in natural environments [13, 27].

Blue-grey and green roofs were also installed at Oslo with the suggested LECA® LWA build-up, based on recommendations from Andenæs et al. and Thodesen et al. [28, 29] that addresses the performance of blue-grey and green roofs in cold climates.

Oslo municipality has developed a 3-step-strategy against all rainwater (not only the extreme events), to maintain a healthy water cycle and increase water resilience of the city (**Figure 16**). The advantage of using LECA® LWA was due to their potential benefits in all steps of this water management strategy (catchment, infiltration, delay, detention, drainage and protection).

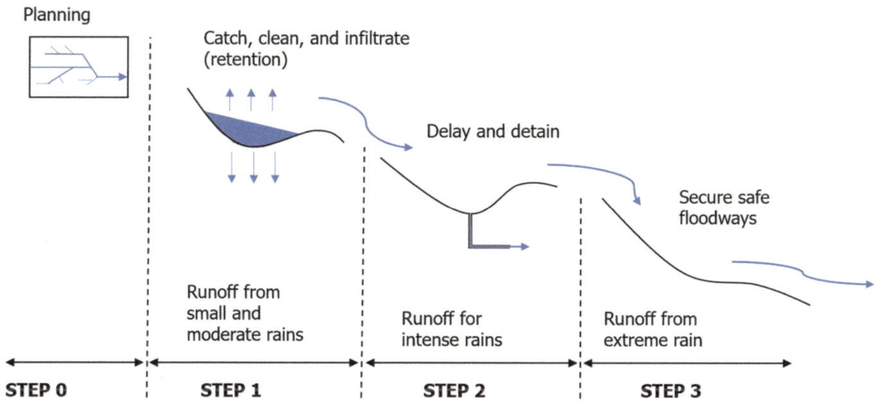

Figure 16.
Oslo 3-step strategy city planning and urban water management.

Project name	Location	Usage/Area	Purpose of LECA® LWA	LECA® LWA type (Volume)	Project photography
Lillebörg Terasse	Stokke, Norway	Rooftop recreational area (450 m^2)	Water detention	Finely crushed LECA® (59 m^3)	
Makrillen	Goteborg, Sweden	Courtyard (2500 m^2)	Sublayer water storage	Round LECA® 10–20 mm (700 m^3)	
Valkyriegata	Oslo, Norway	Rooftop recreational area (320 m^2)	Water detention	Finely crushed LECA® LWA (105 m3)	
Budolfi Plads	Aalborg, Denmark	Public Town square and park (7200 m^2)	Lightweight drainage material	Round LECA® LWA 10–20 (3000 m^3)	
REDI shopping center green deck	Helsinki, Finland	Rooftop recreational area (13.000 m^2)	Terrain shapes, drainage and soil mix	LECA® LWA 4–10 mm and crushed LECA® LWA 3–8 mm (1200 m^3)	
Emporia shopping center	Malmo, Sweden	Rooftop recreational area (27.000 m^2)	Terrain shapes, drainage, insulation	LECA® LWA 10–20 mm (10.000 m^3)	
Drainage ditch, Guimarães national highway	Guimarães, Portugal	Re-profiling of a national highway section	Lightweight drainage material	LECA®(S), LECA®(L) (192 m^3)	
Zielony Nugat II Housing Estate	Warsaw, Poland	Buffer water drainage	Drainage, thermal insulation and substructure	LECA®(L) (300 m^3)	
Laureate Gardens, Henley on Thames	United Kingdom	Rooftop recreational area above a residential car park	Lightweight drainage material	LECA® LWA 10–20 mm (120 m^3)	

Table 4.
LECA® LWA water management reference projects.

3.5 Reference projects for LECA® LWA applications

Stormwater management strategies in a specific location contribute to runoff decrease and lessen water accumulation preventing drainage sewer systems overflow. As such, several other water management reference projects have been constructed taking advantage of LECA® LWA properties and their water management capacity. **Table 4** presents a few examples of such projects.

4. Conclusion

Rainwater management in urban areas is a major concern in the present climate change scenario, with more frequent and intense precipitation events and the consequent urban floods and water resources degradation. NbS have been increasingly implemented within cities, due to their operational benefits in minimising stormwater management problems by catching, retaining, detaining and infiltrating the excess water when intense precipitation events occur. However, NbS performance can be enhanced by adding a natural material—LECA® LWA, as a layer. This lightweight material is applicable to all traditional NbS solutions, and increases the simulation of nature's own water infiltration mechanisms due to its high water retention capacity—thus being ideal for local water management in all types of infrastructures. Experimental studies show that by using LECA® LWA in a green roof sublayer, a reduction on the runoff volume of about 95% has been achieved and a delay in the peak flow at around 1 h 15 min. Permeable pavements with LECA® LWA can also significantly increase their runoff coefficient from 33–89% when compared to the pavement alone. On the other hand, the LECA® LWA granulometry has a significant influence on the NbS. Finely crushed LECA® LWA presents the lower runoff coefficient values then coarsely or round LECA® LWA (0.2–0.5 and 0.3–0.7, respectively), revealing a better retention performance.

Several reference projects have been implemented in Europe specifying LECA® LWA in their sublayers to increase the performance of those structures. Laboratory and field tests have shown that by using a layer of LECA® LWA, a peak runoff intensity reduction has been achieved as well as a peak delay, and also higher amounts of detained water when compared to systems without LECA® LWA layer. However, the water retention capacity of LECA® LWA is limited and when it reaches saturation level, stormwater retention capacity of the system incorporating LWA ceases and the runoff starts flowing to drainage systems. Further investigations are thus necessary to continue scientific research and development of NbS with higher hydraulic efficiency using LECA® LWA with different granulometry in their composition, considering their inherent operating differences and their distinct interaction with the surrounding environment. Additionally, NbS combination with other LID or BMP's solutions must be considered, since it can enhance its stormwater retention capacity and water quality improvement to a level where it can become an alternative water source to urban consumption.

Furthermore, the good planning and development of such NbS as the one implemented in Oslo, can combine recreational areas and stormwater water management measures, thus improving the standard of living in densely populated urban cities, creating at the same time, more pleasant and accessible places. The development of such blue-grey/green solutions are imperative, and will challenge the construction industry norms when planning local water management measures, contributing at the same time to urban areas resilience and climate change mitigation.

Acknowledgements

Acknowledgments: Cristina M. Monteiro and Cristina Santos would like to thank LECA® Portugal the financial support for the publication of the present chapter. Cristina M. Monteiro would also like to thank Fundação para a Ciência e Tecnologia (FCT) under the FCT project UIDB/50016/2020.

Conflict of interest

No conflict of interest.

Author details

Cristina M. Monteiro[1*], Cristina Santos[2,3], Jaran R. Wood[4] and Kim Rosenbom[5]

1 Universidade Católica Portuguesa, CBQF—Centro de Biotecnologia e Química Fina—Laboratório Associado, Escola Superior de Biotecnologia, Porto, Portugal

2 Faculdade de Engenharia da Universidade do Porto, Porto, Portugal

3 CIIMAR—Centro Interdisciplinar de Investigação Marinha e Ambiental, Matosinhos, Portugal

4 Leca International, Norway

5 Leca International, Copenhagen S, Denmark

*Address all correspondence to: cmonteiro@ucp.pt

References

[1] Communication from the commission to the European parliament, the council, the European economic and social committee and the committee of the regions—Forging a climate-resilient Europe—The new EU strategy on adaptation to climate change—European Commission, Brussels 24.2.2021; COM (2021)82 final

[2] Ecosystem-Based Approaches (GI). Available from: https://climate-adapt.eea.europa.eu/eu-adaptation-policy/sector-policies/ecosystem

[3] EEA. Nature-based solutions in Europe: Policy, knowledge and practice for climate change adaptation and disaster risk reduction—EEA report n° 01/2021. EEA, Copenhagen K, Denmark. DOI: 10.2800/919315

[4] Pla C, Benavente D, Valdes-Abellan J, Kovacova Z. Effectiveness of two lightweight aggregates for the removal of heavy metals from contaminated urban stormwater. Journal of Contaminant Hydrology. 2021;**239**:103778. DOI: 10.1016/j.jconhyd.2021.103778

[5] Mlih R, Bydalek F, Klumpp E, Yaghi N, Bol R, Wenk J. Light-expanded clay aggregate (LECA) as a substrate in constructed wetlands—A review. Ecological Engineering. 2020;**148**:105783. DOI: 10.1016/j.ecoleng.2020.105783

[6] Karami H, Teymouri E, Mousavi S-F, Farzin S. Experimental investigation of the effect of adding LECA and pumice on some physical properties of porous concrete. Engineering Journal. 2018;**22**:205-213. DOI: 10.4186/ej.2018.22.1.205

[7] Molineux CJ, Newport DJ, Ayati B, Wang C, Connop SP, Green JE. Bauxite residue (red mud) as a pulverised fuel ash substitute in the manufacture of lightweight aggregate. Journal of Cleaner Production. 2016;**112**:401-408. DOI: 10.1016/j.jclepro.2015.09.024

[8] Pradhan S, Al-Ghamdi SG, Mackey HR. Greywater recycling in buildings using living walls and green roofs: A review of the applicability and challenges. Science of the Total Environment. 2019;**652**:330-344. DOI: 10.1016/j.scitotenv.2018.10.226

[9] Sailor DJ, Hagos M. An updated and expanded set of thermal property data for green roof growing media. Energy Buildings. 2011;**43**:2298-2303. DOI: 10.1016/j.enbuild.2011.05.014

[10] Sengul O, Azizi S, Karaosmanoglu F, Tasdemir MA. Effect of expanded perlite on the mechanical properties and thermal conductivity of lightweight concrete. Energy Buildings. 2011;**43**:671-676. DOI: 10.1016/j.enbuild.2010.11.008

[11] World Bank 2021. A Catalogue-of-Nature-based-Solutions-for-Urban-Resilience. Washington, DC: The World Bank Group; 2021

[12] FEMA. Building community resilience with nature-based solutions—A guide for local communities. 2021.

[13] Jotte L, Raspati G, Azrague K. Review of stormwater management practices. KLIMA 2050, report 7; Publisher: SINTEF Building and Infrastructure, Høgskoleringen 7 b, POBox 4760 Sluppen, N-7465 Trondheim

[14] How to Manage Stormwater | City of Oberlin. Available from: https://www.cityofoberlin.com/city-government/departments/

public-works/stormwater-management/how-to-manage-stormwater/

[15] Natural Water Retention Measures (nwrm.eu). Available from: http://nwrm.eu/measure

[16] Facilities—Infiltration basin|Stormwater Partners. Available from: https://www.stormwaterpartners.com/facilities-infiltration-basin

[17] Usaquén Urban Wetland by Obraestudio|Inhabitat—Green Design, Innovation, Architecture, Green Building. Available from: https://inhabitat.com/an-urban-wetland-springs-to-life-among-bogotas-high-rises/usaquen-urban-wetland-by-obraestudio-3/

[18] The Permeable Paving Guide—Part 1: Introduction to Permeable Pavements|Brickworks. Available from: https://www.brickworks.com.au/articles/permeable-paving-guide-part-1-introduction-permeable-pavements/

[19] LECA® Stormwater Management Technical Guide. How to detain and drain water and rain. Available from: https://www.leca.no/sites/leca.no/files/pdf/Storm%20water%20technical%20guidelines%20brosjyre.pdf

[20] Wood J. Stormwater detention factors for a blue-green roof based on lightweight expanded clay aggregate in Norway. In: Extended Abstract Conference Paper; Nordic Geotechnical Meeting 2020 (NGM 2020). Available from: https://ngm2020.exordo.com/ [Accessed: 26 January 2022]

[21] Pradhan S, Helal MI, Al-Ghamdi SG, Mackey HR. Performance evaluation of various individual and mixed media for greywater treatment in vertical nature-based systems. Chemosphere. 2020;**245**:125564. DOI: 10.1016/j.chemosphere.2019.125564

[22] Lee J, Lee M, Han M. A pilot study to evaluate runoff quantity from green roofs. Journal of Environmental Management. 2015;**152**:171-176

[23] Holt E, Koivusalo H, Korkealaakso J, Sillanpää N, Wendling L. Filtration Systems for Stormwater Quantity and Quality Management: Guideline for Finnish Implementation. VTT Technical Research Centre of Finland. 2018. VTT Technology No. 338

[24] Schärer LA. Comparing experimentally measured runoff coefficients with field observations for detention-based roofs [thesis]. NTNU (Norwegian University of Science and Technology—Department of Civil and Environmental Engineering); 2018

[25] Koc K, Ekmekcioğlu Ö, Özger M. An integrated framework for the comprehensive evaluation of low impact development strategies. Journal of Environmental Management. 2021;**294**:113023. DOI: 10.1016/j.jenvman.2021.113023

[26] Harper G, Limmer M, Showalter E, Burken J. Nine-month evaluation of runoff quality and quantity from an experiential green roof in Missouri, USA. Ecological Engineering. 2015;**78**:127-133. DOI: 10.1016/j.ecoleng.2014.06.004

[27] Hamouz V, Lohne J, Wood JR, Muthanna TM. Hydrological performance of LECA-based roofs in cold climates. Water. 2018;**10**:263-278. DOI: 10.3390/w10030263

[28] Andenæs E, Kvande T, Muthanna TM, Lohne J. Performance of blue-green roofs in cold climates: A scoping review.

Buildings. 2018;**8**:55-78. DOI: 10.3390/buildings8040055

[29] Thodesen B, Kvande T, Tajet HT, Time B, Lohne J. Adpating green-blue roofs to Nordic climate. Nordic Journal of Architectural Research. 2018;**2**:99-126